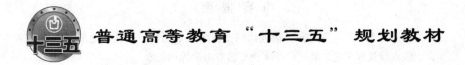

普通高等教育"十三五"规划教材

冶金工程专业英语

炼 钢 篇

Specialized English in Metallurgical Engineering

Steelmaking Section

主　编　孙立根
副主编　李慧蓉　陈　伟　张淑会

北　京
冶金工业出版社
2017

内 容 提 要

本书共5章，主要内容包括现代炼钢技术发展历程、钢铁冶金的基础理论、铁水预处理工艺、氧气转炉炼钢工艺和电炉炼钢技术。

本书以讲授包括铁水预处理在内的炼钢工艺为主，同时介绍了钢铁冶金，特别是炼钢环节涉及的基础理论，以满足不同层次学生学习的需要。

本书可作为高等院校冶金工程等专业的专业英语教材（配有教学课件），所介绍内容能够满足本科生和研究生的教学要求，同时也可供从事相关专业的工程技术人员和管理人员参考。

图书在版编目(CIP)数据

冶金工程专业英语. 炼钢篇/孙立根主编. —北京：冶金工业出版社，2017.8

普通高等教育"十三五"规划教材
ISBN 978-7-5024-7568-0

Ⅰ.①冶… Ⅱ.①孙… Ⅲ.①冶金工业—英语—高等学校—教材 Ⅳ.①TF

中国版本图书馆 CIP 数据核字（2017）第 136436 号

出 版 人　谭学余
地　　址　北京市东城区嵩祝院北巷39号　邮编　100009　电话　(010)64027926
网　　址　www.cnmip.com.cn　电子信箱　yjcbs@cnmip.com.cn
责任编辑　杜婷婷　美术编辑　彭子赫　版式设计　孙跃红
责任校对　郑　娟　责任印制　牛晓波

ISBN 978-7-5024-7568-0

冶金工业出版社出版发行；各地新华书店经销；三河市双峰印刷装订有限公司印刷
2017年8月第1版，2017年8月第1次印刷
787mm×1092mm　1/16；12.5印张；301千字；189页
36.00元

冶金工业出版社　投稿电话　(010)64027932　投稿信箱　tougao@cnmip.com.cn
冶金工业出版社营销中心　电话　(010)64044283　传真　(010)64027893
冶金书店　地址　北京市东四西大街46号(100010)　电话　(010)65289081(兼传真)
冶金工业出版社天猫旗舰店　yjgycbs.tmall.com

(本书如有印装质量问题，本社营销中心负责退换)

前　言

　　炼钢是钢铁冶炼过程的核心模块之一，也是高等院校冶金工程专业学生必须掌握的知识环节。随着我国国民经济的快速发展和国际交流步伐的加快，专业英语已成为学生大学阶段专业知识学习的重要组成部分，但目前专业英语教材建设相对滞后。因此，作者团队依据课程教学大纲，在多年讲授冶金工程专业英语课程和相关领域课程的基础上，参考英文书籍，编写了本书。

　　本书作为《冶金工程专业英语》系列教材之一，重点讲授了包括铁水预处理在内的炼钢工艺，同时还讲解了钢铁冶金，特别是炼钢环节涉及的基础理论。本书可以有效帮助学生提高查阅专业英文文献的阅读效率，同时对提升学生与国外同行专家进行技术交流的表达能力也会有很大帮助。

　　本书共5章。第1章现代炼钢技术发展历程，重点讲述不同类型炼钢工艺的发展历程。第2章钢铁冶金的基础理论，重点介绍了炼钢环节涉及的基本冶金反应原理。第3章铁水预处理工艺，重点介绍了铁水预脱硫工艺。第4章氧气转炉炼钢工艺，围绕转炉操作、原料、炉内反应和能量守恒、冶炼参数控制以及排放物等进行了详细的阐述。第5章电炉炼钢技术，从电炉冶炼工艺出发，阐述了电炉原料、造渣料、电炉操作和废钢熔化工艺等电炉冶炼的关键环节。

　　在本书编写过程中，除编写团队外，郝剑桥、刘云松等研究生为书中图表的整理和编辑做了大量工作，在此表示诚挚的感谢。

　　本书配套教学课件读者可在冶金工业出版社官网（http：//www.cnmip.com.cn）输入书名搜索资源并下载。

　　由于作者水平所限，书中不妥之处，恳请读者批评指正。

作　者
2017 年 5 月

Preface

Steelmaking is one of the core modules for iron and steelmaking process, and it is also the key knowledge which must be mastered by the metallurgical engineering students in colleges and universities. With the rapid development of China's national economy and the big pace of international exchanges, the specialized English has become an important part of the professional knowledge for university students, but at present, the development of specialized English textbooks is lagging behind relatively. According to the course syllabus, referring to the classical English references, and based on the abundant teaching experiences of metallurgical engineering English courses and other related courses, the author team redacted this textbook.

As one of the series of "specialized English in metallurgical engineering" textbooks, the steelmaking technology which is also included the pre-treatment of hot me-tal, and the fundamentals for steelmaking had been focused on in this book. This book can effectively help students improve the reading efficiency of professional English references, and is also helpful to enhance the ability of students to communicate with foreign counterparts.

There are 5 chapters in this book. Chapter 1 is the historical development of modern steelmaking, and focus on the development of different types of steelmaking process. Chapter 2 is the fundamentals of iron and steelmaking, and focus on the basic metallurgical reaction principle involved in steelmaking. Chapter 3 is the pre-treatment of hot metal, and focus on the pre-desulphurization process of molten iron. Chapter 4 is the oxygen steelmaking processes, and focus on the converter operation, raw material, reaction in the furnace and energy conservation, smelting parameter control and emissions control. Chapter 5 is the electric furnace steelmaking, and focus on the electric furnace smelting technology, including raw materials, slagging materials, electric furnace operation and scrap melting process.

In the preparation of this book, beside the author team, the graduate students Hao Jianqiao, Liu Yunsong etc. had done a lot of work for chart finishing and editing in this book, authors would like to express our sincere thanks to them.

The teaching courseware affiliated to this book could be downloaded from the Metallurgical Industry Press official website (http://www.cnmip.com.cn).

For the author's limited knowledge, if there are inappropriate contents in the book, please tell us to correct, thanks.

The authors
May 2017

Contents

1 Historical Development of Modern Steelmaking 1

    1.1 Bottom-Blown Acid or Bessemer Process 1
    1.2 Basic Bessemer or Thomas Process 3
    1.3 Open Hearth Process 4
    1.4 Oxygen Steelmaking 7
    1.5 Electric Furnace Steelmaking 9
    Exercises 11

2 Fundamentals of Iron and Steelmaking 12

    2.1 Fundamentals of Steelmaking Reactions 12
        2.1.1 Slag-Metal Equilibrium in Steelmaking 12
        2.1.2 State of Reactions in Steelmaking 16
    2.2 Fundamentals of Reactions in Electric Furnace Steelmaking 26
        2.2.1 Slag Chemistry and the Carbon, Manganese, Sulfur and Phosphorus Reactions in the EAF 26
        2.2.2 Control of Residuals in EAF Steelmaking 28
        2.2.3 Nitrogen Control in EAF Steelmaking 29
    2.3 Fundamentals of Stainless Steel Production 30
        2.3.1 Decarburization of Stainless Steel 30
        2.3.2 Nitrogen Control in the AOD 32
        2.3.3 Reduction of Cr from Slag 33
    2.4 Fundamentals of Ladle Metallurgical Reactions 34
        2.4.1 Deoxidation Equilibrium and Kinetics 34
        2.4.2 Ladle Desulfurization 41
        2.4.3 Calcium Treatment of Steel 43
    2.5 Fundamentals of Degassing 44
        2.5.1 Fundamental Thermodynamics 44
        2.5.2 Vacuum Degassing Kinetics 45
    Exercises 47

3 Pre-treatment of Hot Metal 48

    3.1 Introduction 48

3.2　Desiliconization and Dephosphorization Technologies ··················· 49
3.3　Desulfurization Technology ··················· 52
　3.3.1　Introduction ··················· 52
　3.3.2　Process Chemistry ··················· 53
　3.3.3　Transport Systems ··················· 57
　3.3.4　Process Venue ··················· 58
　3.3.5　Slag Management ··················· 59
　3.3.6　Lance Systems ··················· 59
　3.3.7　Cycle Time ··················· 61
　3.3.8　Hot Metal Sampling and Analysis ··················· 61
　3.3.9　Reagent Consumption ··················· 62
　3.3.10　Economics ··················· 62
　3.3.11　Process Control ··················· 62
3.4　Hot Metal Thermal Adjustment ··················· 63
Exercises ··················· 64

4　Oxygen Steelmaking Processes ··················· 65

4.1　Introduction ··················· 65
　4.1.1　Process Description and Events ··················· 65
　4.1.2　Types of Oxygen Steelmaking Processes ··················· 67
　4.1.3　Environmental Issues ··················· 68
4.2　Sequence of Operations—Top Blown ··················· 68
　4.2.1　Plant Layout ··················· 68
　4.2.2　Sequence of Operations ··················· 70
　4.2.3　Shop Manning ··················· 76
4.3　Raw Materials ··················· 80
　4.3.1　Introduction ··················· 80
　4.3.2　Hot Metal ··················· 80
　4.3.3　Scrap ··················· 82
　4.3.4　High Metallic Alternative Feeds ··················· 83
　4.3.5　Oxide Additions ··················· 84
　4.3.6　Fluxes ··················· 85
　4.3.7　Oxygen ··················· 87
4.4　Process Reactions and Energy Balance ··················· 88
　4.4.1　Reactions in BOF Steelmaking ··················· 88
　4.4.2　Slag Formation in BOF Steelmaking ··················· 91
　4.4.3　Mass and Energy Balances ··················· 92
　4.4.4　Tapping Practices and Ladle Additions ··················· 96
4.5　Process Variations ··················· 97

  4.5.1 The Bottom-Blown Oxygen Steelmaking or OBM (Q-BOP) Process ········ 97
  4.5.2 Mixed-Blowing Processes ·· 102
  4.5.3 Oxygen Steelmaking Practice Variations ································ 105
 4.6 Process Control Strategies ·· 108
  4.6.1 Introduction ·· 108
  4.6.2 Static Models ·· 109
  4.6.3 Statistical and Neural Network Models ································ 111
  4.6.4 Dynamic Control Schemes ·· 112
  4.6.5 Lance Height Control ·· 114
 4.7 Environmental Issues ·· 114
  4.7.1 Basic Concerns ·· 114
  4.7.2 Sources of Air Pollution ·· 115
  4.7.3 Relative Amounts of Fumes Generated ································ 117
  4.7.4 Other Pollution Sources ·· 118
  4.7.5 Summary ··· 119
 Exercises ·· 119

5 Electric Furnace Steelmaking ·· 121

 5.1 Electric Furnace Technology ·· 121
  5.1.1 Oxygen Use in the EAF ·· 122
  5.1.2 Oxy-Fuel Burner Application in the EAF ····························· 122
  5.1.3 Application of Oxygen Lancing in the EAF ·························· 125
  5.1.4 Foamy Slag Practice ··· 128
  5.1.5 CO Post-Combustion ·· 130
  5.1.6 EAF Bottom Stirring ··· 140
  5.1.7 Furnace Electrics ··· 140
  5.1.8 High Voltage AC Operations ··· 142
  5.1.9 DC EAF Operations ··· 143
  5.1.10 Use of Alternative Iron Sources in the EAF ························ 146
  5.1.11 Conclusions ·· 147
 5.2 Raw Materials ··· 148
 5.3 Fluxes and Additives ·· 149
 5.4 Furnace Operations ·· 151
  5.4.1 EAF Operating Cycle ·· 151
  5.4.2 Furnace Charging ··· 152
  5.4.3 Melting ··· 153
  5.4.4 Refining ·· 154
  5.4.5 Deslagging ··· 156
  5.4.6 Tapping ·· 157

5.4.7 Furnace Turnaround 157
5.4.8 Furnace Heat Balance 158
5.5 New Scrap Melting Processes 159
5.5.1 Scrap Preheating 159
5.5.2 Preheating with Offgas 160
5.5.3 Natural Gas Scrap Preheating 161
5.5.4 K-ES 161
5.5.5 Danarc Process 164
5.5.6 Fuchs Shaft Furnace 165
5.5.7 Consteel Process 173
5.5.8 Twin Shell Electric Arc Furnace 176
5.5.9 Processes under Development 179
Exercises 187

References 189

1 Historical Development of Modern Steelmaking

1.1 Bottom-Blown Acid or Bessemer Process

This process, developed independently by William Kelly of Eddyville, Kentucky and Henry Bessemer of England, involved blowing air through a bath of molten pig iron contained in a bottom-blown vessel lined with acid (siliceous) refractories. The process was the first to provide a large scale method whereby pig iron could rapidly and cheaply be refined and converted into liquid steel. Bessemer's American patent was issued in 1856; although Kelly did not apply for a patent until 1857, he was able to prove that he had worked on the idea as early as 1847. Thus, both men held rights to the process in this country; this led to considerable litigation and delay, as discussed later. Lacking financial means, Kelly was unable to perfect his invention and Bessemer, in the face of great difficulties and many failures, developed the process to a high degree of perfection and it came to be known as the acid Bessemer process.

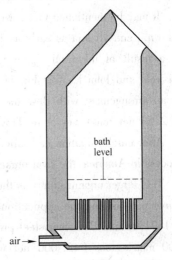

Fig. 1.1 Principle of the bottom blown converter. The blast enters the wind box beneath the vessel through the pipe indicated by the arrow and passes into the vessel through tuyeres set in the bottom of the converter

The fundamental principle proposed by Bessemer and Kelly was that the oxidation of the major impurities in liquid blast furnace iron (silicon, manganese and carbon) was preferential and occurred before the major oxidation of iron, and the actual mechanism differs from this simple explanation. Further, they discovered that sufficient heat was generated in the vessel by the chemical oxidation of the above elements in most types of pig iron to permit the simple blowing of cold air through molten pig iron to produce liquid steel without the need for an external source of heat. Because the process converted pig iron to steel, the vessel in which the operation was carried out came to be known as a converter. The principle of the bottom blown converter is shown schematically in Fig. 1.1.

At first, Bessemer produced satisfactory steel in a converter lined with siliceous (acid) refractories by refining pig iron that, smelted from Swedish ores, was low in phosphorus, high in manganese, and contained enough silicon to meet the thermal needs of the process. But, when applied to irons which were higher in phosphorus and low in silicon and manganese, the process did not produce satisfactory steel. In order to save his process in the face of opposition among steelmakers, Bessemer built a steel works at Sheffield, England, and began to operate in 1860. Even when low phosphorus Swedish pig iron was employed, the steels first produced there contained much more than the admissible amounts of oxygen, which made the steel "wild" in the molds. Difficulty also was experienced with sulfur, introduced from the coke used as the fuel for melting the iron in cupolas, which contributed to "hot shortness" of the steel. These objections finally were overcome by the addition of manganese in the form of spiegeleisen to the steel after blowing as completed.

The beneficial effects of manganese were disclosed in a patent by R. Mushet in 1856. The carbon and manganese in the spiegeleisen served the purpose of partially deoxidizing the steel, which part of the manganese combined chemically with some of the sulfur to form compounds that either floated out of the metal into the slag, or were comparatively harmless if they remained in the steel.

As stated earlier, Bessemer had obtained patents in England and in this country previous to Kelly's application. Therefore, both men held rights to the process in the United States.

The Kelly Pneumatic Process Company had been formed in 1863 in an arrangement with William Kelly for the commercial production of steel by the new process. This association included the Cambria Iron Company, E. B. Ward, Park Brothers and Company, Lyon, Shord and Company, Z. S. Durfee and , later, Chouteau, Harrison and Vale. This company, in 1864, built the first commercial Bessemer plant in this country, consisting of a 2.25 metric tons (2.50 net tons) acid lined vessel erected at the Wyandotte Iron Works, Wyandotte, Michigan, owned by Captain E. B. Ward. It may be mentioned that a Kelly converter was used experimentally at the Cambria Works, Johnstown, Pennsylvania as early as 1861.

As a result of the dual rights to the process a second group consisting of Messrs. John A. Griswold and John F. Winslow of Troy, New York and A. L. Holley formed another company under an arrangement with Bessemer in 1864. This group erected an experimental 2.25 metric tons (2.50 net tons) vessel in Troy, New York which commenced operations on February 16, 1865. After much litigation had failed to gain for either sole control of the patents for the pneumatic process in America, the rival organizations decided to combine their respective interests early in 1866. This larger organization was then able to combine the best features covered by the Kelly and Bessemer patents, and the application of the process advanced rapidly.

By 1871, annual Bessemer steel production in the United States had increased to approximately 40,800 metric tons (45,000 net tons), about 55% of the total steel production, which was produced by seven Bessemer plants.

Bessemer steel production in the United States over an extended period of years remained significant. However, raw steel is no longer being produced by the acid Bessemer process in the United States. The last completely new plant for the production of acid Bessemer steel ingots in the United States was built in 1949.

As already stated, the bottom blown acid process known generally as the Bessemer Process was the original pneumatic steelmaking process. Many millions of tons of steel were produced by this method.

From 1870 to 1910, the acid Bessemer process produced the majority of the world's supply of steel.

The success of acid Bessemer steelmaking was dependent upon the quality of pig iron available which, in turn, demanded reliable supplies of iron ore and metallurgical coke of relatively high purity. At the time of the invention of the process, large quantities of suitable ores were available, both abroad and in the United States. With the gradual depletion of high quality ores abroad (particularly low phosphorus ores) and the rapid expansion of the use of the bottom blown basic pneumatic, basic open hearth and basic oxygen steelmaking processes over the years, acid Bessemer steel production has essentially ceased in the United Kingdom and Europe.

In the United States, the Mesabi Range provided a source of relatively high grade ore for making iron for the acid Bessemer process for many years. In spite of this, the acid Bessemer process declined from a major to a minor steelmaking method in the United States and eventually was abandoned.

The early use of acid Bessemer steel in this country involved production of a considerable quantity of rail steel, and for many years (from its introduction in 1864 until 1908) this process was the principal steelmaking process. Until relatively recently, the acid Bessemer process was used principally in the production of steel for buttwelded pipe, seamless pipe, free machining bars, flat rolled products, wire, steel castings, and blown metal for the duplex process.

Fully killed acid Bessemer steel was used for the first time commercially by United States Steel Corporation in the production of seamless pipe. In addition, dephosphorized acid Bessemer steel was used extensively in the production of welded pipe and galvanized sheets.

1.2 Basic Bessemer or Thomas Process

The bottom blown basic pneumatic process, known by several names including Thomas, Thomas-Gilchrist or basic Bessemer process, was patented in 1879 by Sidney G. Thomas in England. The process, involving the use of the basic lining and a basic flux in the converter, made it possible to use the pneumatic method for refining pig irons smelted from the high phosphorus ores common to many sections of Europe. The process (never adopted in the United States) developed much more rapidly in Europe than in Great Britain and, in 1890, European production was over 1.8 million metric tons (2 million net tons) as compared with 0.36 million metric tons (400,000 net tons) made in Great Britain.

The simultaneous development of the basic open hearth process resulted in a decline of production of steel by the bottom blown basic pneumatic process in Europe and, by 1904, production of basic open hearth steel there exceeded that of basic pneumatic steel. From 1910 on, the bottom blown basic pneumatic process declined more or less continuously percentage-wise except for the period covering World War II, after which the decline resumed.

1.3 Open Hearth Process

Karl Wilhelm Siemens, by 1868, proved that it was possible to oxidize the carbon in liquid pig iron using iron ore, the process was initially known as the "pig and ore process". Briefly, the method of Siemens was as follows. A rectangular covered hearth was used to contain the charge of pig iron or pig iron and scrap (Fig. 1.2). Most of the heat required to promote the chemical reactions necessary for purification of the charge was provided by passing burning fuel gas over the top of the materials. The fuel gas, with a quantity of air more than sufficient to burn it, was introduced through ports at each end of the furnace, alternately at one end and then the other. The products of combustion passed out of the port temporarily not used for entrance of gas and air, and entered chambers partly filled with brick checkerwork. This checkerwork, commonly called checkers, provided a multitude of passageways for the exit of the gases to the stack. During their passage through the checkers, the gases gave up a large part of their heat to the brickwork. After a short time, the gas and air were shut off at the one end and introduced into the furnace through the preheated checkers, absorbing some of the heat stored in these checkers. The gas and air were thus preheated to a

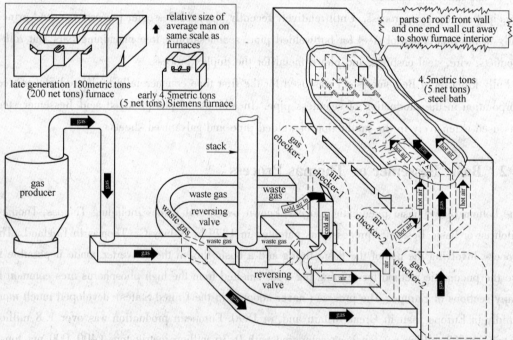

Fig. 1.2 Schematic arrangement of an early type of Siemens furnace with about a 4.5 metric tons (5 net tons) capacity

The roof of this design (which was soon abandoned) dipped from the ends toward the center of the furnace to force the flame downward on the bath. Various different arrangements of gas and air ports were used in later furnaces.

Note that in this design, the furnace proper was supported on the regenerator arches. Flow of gas, air and waste gases were reversed by changing the position of the two reversing valves. The inset at the upper left compares the size of one of these early furnaces with that of a late generation 180 metric tons (200 net tons) open hearth.

somewhat elevated temperature, and consequently developed to a higher temperature in combustion than could be obtained without preheating. In about 20 minutes, the flow of the gas and air was again reversed so that they entered the furnace through the checkers and port used first; and a series of such reversals, occurring every fifteen or twenty minutes was continued until the heat was finished. The elements in the bath which were oxidized both by the oxygen of the air in the furnace atmosphere and that contained in the iron ore fed to the bath, were carbon, silicon and manganese, all three of which could be reduced to as low a limit as was possible in the Bessemer process. Of course, a small amount of iron remains or is oxidized and enters the slag.

Thus, as in all other processes for purifying pig iron, the basic principle of the Siemens process was that of oxidation. However, in other respects, it was unlike any other process. True, it resembled the puddling process in both the method and the agencies employed, but the high temperatures attainable in the Siemens furnace made it possible to keep the final product molten and free of entrapped slag. The same primary result was obtained as in the Bessemer process, but by a different method and through different agencies, both of which imparted to steel made by the new process properties somewhat different from Bessemer steel, and gave the process itself certain metallurgical advantages over the older pneumatic process, as discussed later in this section.

As would be expected, many variations of the process, both mechanical and metallurgical, have been worked out since its original conception. Along mechanical lines, various improvements in the design, the size and the arrangement of the parts of the furnace have been made. Early furnaces had capacities of only about 3.5-4.5 metric tons (4-5 net tons), which modern furnaces range from about 35-544 metric tons (40-600 net tons) in capacity, with the majority having capacities between about 180-270 metric tons (200-300 net tons).

The Siemens process became known more generally, as least in the United States, as the open hearth process. The name "open hearth" was derived, probably, from the fact that the steel, while melted on a hearth under a roof, was accessible through the furnace doors for inspection, sampling and testing.

The hearth of Siemens' furnace was of acid brick construction, on top of which the bottom was made up of sand, essentially as in the acid process of today. Later, to permit the charging of limestone and use of a basic slag for removal of phosphorus, the hearth was constructed with a lining of magnesite brick, covered with a layer of burned dolomite or magnesite, replacing the siliceous bottom of the acid furnace. These furnaces, therefore, were designated as basic furnaces, and the process carried out in them was called the basic process. The pig and scrap process was originated by the Martin brothers, in France, who, by substituting scrap for the ore in Siemens' pig and ore process, found it possible to dilute the change with steel scrap to such an extent that less oxidation was necessary.

The advantages offered by the Siemens process may be summarized briefly as follows:

1. By the use of iron ore as an oxidizing agent and by the external application of heat, the temperature of the bath was made independent of the purifying reactions, and the elimination of impurities could be made to take place gradually, so that both the temperature and composition of the

bath were under much better control than in the Bessemer process.

2. For the same reasons, a greater variety of raw materials could be used (particularly scrap, not greatly consumable in the Bessemer converter) and a greater variety of products could be made by the open hearth process than by the Bessemer process.

3. A very important advantage was the increased yield of finished steel from a given quantity of pig iron as compared to the Bessemer process, because of lower inherent sources of iron loss in the former, as well as because of recovery of the iron content of the ore used for oxidation in the open hearth.

4. Finally, with the development of the basic open hearth process, the greatest advantage of Siemens' over the acid Bessemer process was made apparent, as the basic open hearth process is capable of eliminating phosphorus from the bath. While this element can be removed also in the basic Bessemer (Thomas-Gilchrist) process, it is to be noted that, due to the different temperature conditions, phosphorus is eliminated before carbon in the basic open hearth process, whereas the major proportion of phosphorus is not oxidized in the basic Bessemer process until after carbon in the period termed the afterblow. Hence, while the basic Bessemer process requires a pig iron with a phosphorus content of 2.0% or more in order to maintain the temperature high enough for the afterblow, the basic open hearth process permits the economical use of iron of any phosphorus content up to 1.0%. In the United States, this fact was of importance since it made available immense iron ore deposits which could not be utilized otherwise because of their phosphorus content, which was too high to permit their use in the acid Bessemer or acid open hearth process and too low to use in the basic Bessemer process.

The open hearth process became the dominant process in the United States. As early as 1868, a small open hearth furnace was built at Trenton, New Jersey, but satisfactory steel at a reasonable cost did not result and the furnace was abandoned. Later, at Boston, Massachusetts, a successful furnace was designed and operated, beginning in 1870. Following this success, similar furnaces were built at Nashua, New Hampshire and in Pittsburgh, Pennsylvania, the latter by Singer, Nimick and Company, in 1871. The Otis Iron and Steel Company constructed two 6.3 metric tons (7 net tons) furnaces at their Lakeside plant at Cleveland, Ohio in 1874. Two 13.5 metric tons (15 net tons) furnaces were added to this plant in 1878, two more of the same size in 1881, and two more in 1887. All of these furnaces had acid linings, using a sand bottom for the hearths.

The commercial production of steel by the basic process was achieved first at Homestead, Pennsylvania. The initial heat was tapped March 28, 1888. By the close of 1890, there were 16 basic open hearth furnaces operating. From 1890 to 1900, magnesite for the bottom began to be imported regularly and the manufacture of silica refractories for the roof was begun in American plants. For these last two reasons, the construction of basic furnaces advanced rapidly and, by 1900, furnaces larger than 45 metric tons (50 net tons) were being planned.

While the Bessemer process could produce steel at a possibly lower cost above the cost of materials, it was restricted to ores of a limited phosphorus content and its use of scrap was also limited.

The open hearth was not subject to these restrictions, so that the annual production of steel by the open hearth process increased rapidly, and in 1908, passed the total tonnage produced yearly by the Bessemer process. Total annual production of Bessemer steel ingots decreased rather steadily after 1908, and has ceased entirely in the United States. In addition to the ability of the basic open hearth furnace to utilize irons made from American ores, as discussed earlier, the main reasons for proliferation of the open hearth process were its ability to produce steels of many compositions and its ability to use a large proportion of iron and steel scrap, if necessary. Also steels made by any of the pneumatic processes that utilize air for blowing contain more nitrogen than open hearth steels; this higher nitrogen content made Bessemer steel less desirable than open hearth steel in some important applications.

With the advent of oxygen steelmaking which could produce steel in a fraction of the time required by the open hearth process, open hearth steelmaking has been completely phased out in the United States. The last open hearth meltshop closed at Geneva Steel Corporation at Provo, Utah in 1991. Worldwide there are only a relative few open hearths still producing steel.

1.4 Oxygen Steelmaking

Oxygen steelmaking has become the dominant method of producing steel from blast furnace hot metal. Although the use of gaseous oxygen (rather than air) as the agent for refining molten pig iron and scrap mixtures to produce steel by pneumatic processes received the attention of numerous investigators from Bessemer onward, it was not until after World War II that commercial success was attained.

Blowing with oxygen was investigated by R. Durrer and C. V. Schwarz in Germany and by Durrer and Hellbrugge in Switzerland. Bottom-blown basic lined vessels of the designs they used proved unsuitable because the high temperature attained caused rapid deterioration of the refractory tuyere bottom; blowing pressurized oxygen downwardly against the top surface of the molten metal bath, however, was found to convert the charge to steel with a high degree of thermal and chemical efficiency.

Plants utilizing top blowing with oxygen have been in operation since 1952-1953 at Linz and Donawitz in Austria. These operations, sometimes referred to as the Linz-Donawitz or L-D process were designed to employ pig iron produced from local ores that are high in manganese and low in phosphorus; such iron is not suitable for either the acid or basic bottom blown pneumatic process utilizing air for blowing. The top blown process, however, is adapted readily to the processing of blast furnace metal of medium and high phosphorus contents and is particularly attractive where it is desirable to employ a steelmaking process requiring large amounts of hot metal as the principal source of metallics. This adaptability has led to the development of numerous variations in application of the top-blown principle. In its most widely used form, which also is the form used in the United States, the top blown oxygen process is called the basic oxygen steelmaking process (BOF for short) or in some companies the basic oxygen process (BOP for short).

The basic oxygen process consists essentially of blowing oxygen of high purity onto the surface of the bath in a basic lined vessel by a water cooled vertical pipe or lance inserted through the mouth of the vessel (Fig. 1.3).

A successful bottom blown oxygen steelmaking process was developed in the 1970s. Based on development in Germany and Canada and known as the OBM process, or Q-BOP in the United States, the new method has eliminated the problem of rapid bottom deterioration encountered in earlier attempts to bottom blow with oxygen. The tuyeres (Fig. 1.4), mounted in a removable bottom, are designed in such a way that the stream of gaseous oxygen passing through a tuyere into the vessel is surrounded by a sheath of another gas. The sheathing gas is normally a hydrocarbon gas such as propane or natural gas. Vessel capacities of 200 tons and over, comparable to the capacities of typical top blown BOF vessels, are commonly used.

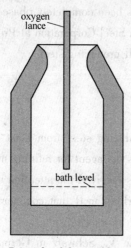

Fig. 1.3 Principle of the top blown converter
Oxygen of commercial purity, at high pressure and velocity is blown downward vertically into surface of bath through a single water cooled pipe or lance.

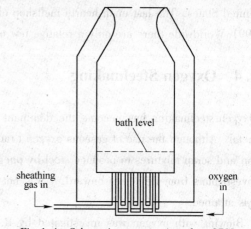

Fig. 1.4 Schematic cross-section of an OBM (Q-BOP) vessel, showing how a suitable gas is introduced into the tuyeres to completely surround the stream of gaseous oxygen passing through the tuyeres into the molten metal bath

The desire to improve control of the oxygen pneumatic steelmaking process has led to the development of various combination blowing processes. In these processes, 60%-100% of the oxygen required to refine the steel is blown through a top mounted lance (as in the conventional BOF) while additional gas (such as oxygen, argon, nitrogen, carbon dioxide or air) is blown through bottom mounted tuyeres or permeable brick elements. The bottom blown gas results in improved mixing of the metal bath, the degree of bath mixing increasing with increasing bottom gas flow rate. By varying the type and flow rate of the bottom gas, both during and after the oxygen blow, specific metallurgical reactions can be controlled to attain desired steel compositions and temperatures. There are, at present many different combination blowing processes, which differ in the type of bottom gas used, the flow rates of bottom gas that can be attained, and the equipment used to introduce the bottom gas into the furnace. All of the processes, to some degree, have similar advantages. The existing combination blowing furnaces are converted conventional BOF furnaces and range in capacity from about 60 tons to more than 300 tons. The conversion to combination blowing began in the late 1970s and has continued at an accelerated rate.

Two other oxygen blown steelmaking, the Stora-Kaldo process and the Rotor process, did not gain wide acceptance.

1.5 Electric Furnace Steelmaking

In the past twenty years there has been a significant growth in electric arc furnace (EAF) steelmaking. When oxygen steelmaking began replacing open hearth steelmaking excess scrap became available at low cost because the BOF melts less scrap than an open hearth. Also for fully developed countries like the United States, Europe and Japan the amount of obsolete scrap in relationship to the amount of steel required increased, again reducing the price of scrap relative to that of hot metal produced from ore and coal. This economic opportunity arising from low cost scrap and the lower capital cost of an EAF compared to integrated steel production lead to the growth of the mini-mill or scrap based EAF producer. At first the mini-mills produced lower quality long products such as reinforcing bars and simple construction materials. However, with the advent of thin slab casting a second generation of EAF plants has developed which produce flat products. In the decade of the 1990's approximately 15-20 million tons of new EAF capacity has been built or planned in North America alone. The EAF has evolved and improved its efficiency tremendously, large quantities of scrap substitutes such as direct reduced iron and pig iron are now introduced in the EAF as well as large quantities of oxygen.

It has been said that arc-type furnaces had their beginning in the discovery of the carbon arc by Sir Humphrey Davy in 1800, but it is more proper to say that their practical application began with the work of Sir William Siemens, who in 1878 constructed, operated and patented furnaces operating on both the direct arc and indirect arc principles.

At this early date, the availability of electric power was limited and its cost high; also, carbon electrodes of the quality required to carry sufficient current for steel melting had not been developed. Thus the development of the electric melting furnace awaited the expansion of the electric power industry and improvement in carbon electrodes.

The first successful commercial EAF was a direct arc steelmaking furnace which was placed in operation by Heroult in 1899. The Heroult patent stated in simple terms, covered single-phase or multi-phase furnaces with the arcs in series through the metal bath. This type of furnace, utilizing three phase power, has been the most successful of the electric furnaces in the production of steel.

In the United States there were no developments along arc furnace lines until the first Heroult furnace was installed in the plant of the Halcomb Steel Company, Syracuse New York, which made its first heat on April 5, 1906. This was a single phase, two electrode, rectangular furnace of 3.6 metric tons (4 net tons) capacity. Two years later a similar but smaller furnace was installed at the Firth-Sterling Steel Company, McKeesport, Pennsylvania, and in 1909, a 13.5 metric tons (15 net tons) three phase furnace was installed in the South Works of the Illinois Steel Company. The latter was, at that time, the largest electric steelmaking furnace in the world, and was the first round (instead of rectangular) furnace. It operated on 25-cycle power at 2200 volts and tapped its first heat on May 10, 1909.

From 1910 to 1980 nearly all the steelmaking EAFs built had three phase alternating current (AC) systems. In the 1980s single electrode direct current (DC) systems demonstrated some advantages over the conventional AC furnaces. In the past 15 years a large percentage of the new EAFs built were DC. Commercial furnaces vary in size from 10 tons to over 300 tons. A typical state-of-the art furnace is 150-180 tons, has several natural gas burners, uses considerable oxygen ($30m^3$/ton), has eccentric bottom tapping and often is equipped with scrap preheating. A schematic of a typical AC furnace is shown in Fig. 1.5.

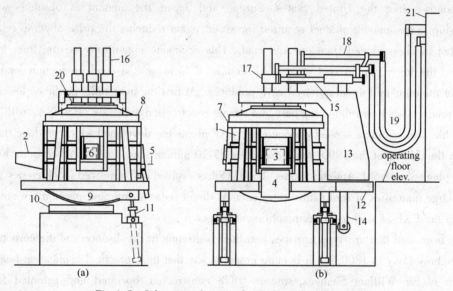

Fig. 1.5 Schematic of a typical AC electric arc furnace

(a) side door elevation; (b) rear door elevation

1—shell; 2—pouring spout; 3—rear door; 4—slag apron; 5—sill line; 6—side door; 7—bezel ring;
8—roof ring; 9—rocker; 10—rocker rail; 11—tilt cylinder; 12—main (tilting) platform;
13—roof removal jib structure; 14—electrode mast stem; 15—electrode mast arm;
16—electrode; 17—electrode holder; 18—bus tube; 19—secondary power cables;
20—electrode gland; 21—electrical equipment vault

Another type of electric melting furnace, used to a certain extent for melting high-grade alloys, is the high frequency coreless induction furnace which gradually replaced the crucible process in the production of complex, high quality alloys used as tool steels. It is used also for remelting scrap from fine steels produced in arc furnaces, melting chrome-nickel alloys, and high manganese scrap, and, more recently, has been applied to vacuum steelmaking processes.

The induction furnace had its inception abroad and first was patented by Ferranti in Italy in 1877. This was a low frequency furnace. It had no commercial application until Kjellin installed and operated one in Sweden. The first large installation of this type was made in 1914 at the plant of the American Iron and Steel Company in Lebanon, Pennsylvania, but was not successful. Low frequency furnaces have operated successfully, especially in making stainless steel.

A successful development using higher frequency current is the coreless high frequency induction furnace. The first coreless induction furnaces were built and installed by the Ajax Electrothermic Corporation, who also initiated the original researches by E. F. Northrup leading to the development of the furnace. For this reason, the furnace is often referred to as the Ajax-Northrup furnace.

The first coreless induction furnaces for the production of steel on a commercial scale were installed at Sheffield, England, and began the regular production of steel in October, 1927. The first commercial steel furnaces of this type in the United States were installed by the Heppenstall Forge and Knife Company, Pittsburgh, Pennsylvania, and were producing steel regularly in November, 1928. Each furnace had a capacity of 272 kilograms (600 pounds) and was served by a 150kV · A motor-generator set transforming 60 hertz current to 860 hertz.

Electric furnace steelmaking has improved significantly in the past twenty years. The tap to tap time, or time required to produce steel, has decreased from about 200 minutes to as little as 55 minutes, electrical consumption has decreased from over 600kW · h per ton to less than 400 and electrical consumption has been reduced by 70%. These have been the result of a large number of technical developments including ultra high power furnaces, long arc practices using foamy slags, the increased use of oxygen and secondary refining. With new EAF plants using scrap alternatives to supplement the scrap charge and the production of higher quality steels, EAF production may exceed 50% in the United States and 40% in Europe and Japan by the year 2010.

Electric furnaces of other various types have been used in the production of steel. These include vacuum arc remelting furnaces (VAR), iron smelting furnaces and on an experimental basis plasma type melting and reheating furnaces. Where appropriate these are discussed in detail in this volume.

Exercises

1-1 How many kinds of processes for steelmaking?

1-2 What is the difference between acid process and basic process?

1-3 What about the advantages of the Siemens process?

1-4 What is the difference between AC and DC for electric furnace steelmaking?

2 Fundamentals of Iron and Steelmaking

There have been tremendous improvements in iron and steelmaking processes in the past thirty years. Productivity and coke rates in the blast furnace and the ability to refine steel to demanding specifications have been improved significantly. Much of this improvement is based on the application of fundamental principles and thermodynamic and kinetic parameters which have been determined. Whereas, many future improvements will be forthcoming in steelmaking equipment, process improvements resulting from the application of fundamental principles and data will likewise continue.

In this chapter the basic principles of thermodynamics and kinetics are reviewed and the relevant thermodynamic data and properties of gases, metals and slags relevant to steelmaking are presented. These principles and data are then applied to steelmaking and secondary refining processes.

In writing this chapter, an attempt has been made to limit the discussion to an average level suitable for the students of metallurgy pursuing graduate or post-graduate education as well as for those with some scientific background engaged in the steel industry. It is assumed that the reader has some basic knowledge of chemistry, physics and mathematics, so that the chapter can be devoted solely to the discussion of the chemistry of the processes.

2.1 Fundamentals of Steelmaking Reactions

In this section of Chapter 2, the discussion of steelmaking reactions will be confined to an assessment of the reaction mechanisms and the state of slag-metal reactions at the time of furnace tapping.

2.1.1 Slag-Metal Equilibrium in Steelmaking

The reaction equilibria in the liquid steel-slag systems have been extensively studied in professional foundation courses, both experimentally and theoretically by applying the principles of thermodynamics and physical chemistry. In a recent reassessment of the available experimental data on steel-slag reactions, it became evident that the equilibrium constants of slag-metal reactions vary with the slag composition in different ways, depending on the type of reaction. For some reactions the slag basicity is the key parameter to be considered; for another reaction the key parameter could be the mass concentration of either the acidic or basic oxide components of the slag.

2.1.1.1 Oxidation of Iron

In steelmaking slags, the total number of g-mols of oxides per 100g of slag is within the range 1.65 ± 0.05. Therefore, the analysis of the slag-metal equilibrium data, in terms of the activity and mol fraction of iron oxide, can be transposed to a simple relation between the mass ratio $[ppm\ O]/(\%FeO)$ and the sum of the acidic oxides $\%SiO_2 + 0.84 \times \%P_2O_5$ as depicted in Fig. 2.1(a). There is of course a corollary relation between the ratio $[ppm\ O]/(\%FeO)$ and the slag basicity as shown in Fig. 2.1(b).

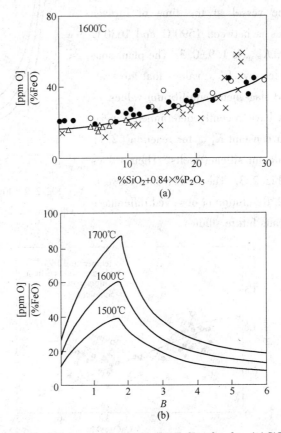

Fig. 2.1 Equilibrium ratio $[ppm\ O]/(\%FeO)$ related to (a) SiO_2 and P_2O_5 contents and (b) slag basicity; experimental data

2.1.1.2 Oxidation of Manganese

For the FeO and MnO exchange reaction involving the oxidation of manganese in steel, formulated below,

$$(FeO) + [Mn] \rightleftharpoons (MnO) + [Fe] \tag{2.1}$$

the equilibrium relation may be described in terms of the mass concentrations of oxides

$$K'_{FeMn} = \frac{(\%MnO)}{(\%FeO)[\%Mn]} \tag{2.2}$$

where (%MnO) ——MnO mass content in the slag;

(%FeO) ——FeO mass content in the slag;

[%Mn] ——Mn mass content in the molten steel.

where the equilibrium relation K'_{FeMn} depends on temperature and slag composition.

The values of K'_{FeMn} derived from the equilibrium constant for reaction (2.1) are plotted in Fig. 2.2 against the slag basicity. In BOF, OBM(QBOP) and EAF steelmaking, the slag basicities are usually in the range 2.5 to 4.0 and the melt temperature in the vessel at the time of furnace tapping in most practices is between 1590℃ and 1630℃ for which the equilibrium K'_{FeMn} is 1.9±0.3. The plant analytical data for tap samples give K'_{FeMn} values that are scattered about the indicated slag-metal equilibrium values.

Morales and Fruehan have recently determined experimentally the equilibrium constant K'_{FeMn} for reaction (2.1) using MgO-saturated calcium silicate melts. Their values of K'_{FeMn} are plotted in Fig. 2.3. The broken line curve is reproduced from Fig. 2.2. Resolution of observed differences in the values of K'_{FeMn} awaits future studies.

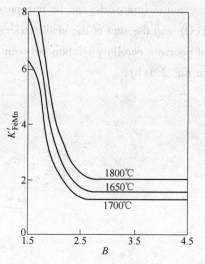

Fig. 2.2 Equilibrium relation in equation (2.2) related to slag basicity

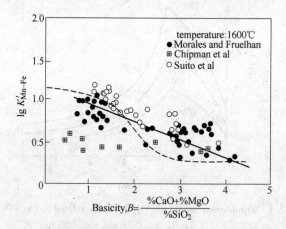

Fig. 2.3 Experimental values of K'_{FeMn} measured recently by Morales and Fruehan

2.1.1.3 Oxidation of Carbon

With respect to the slag-metal reaction, the equilibrium relation for carbon oxidation would be

$$(FeO) + [C] = CO + [Fe] \tag{2.3}$$

$$K_{FC} = \frac{P_{CO}}{[\%C] a_{FeO}}$$

$$\lg K_{FC} = -\frac{5730}{T} + 5.096 \tag{2.4}$$

For 1600℃, $\gamma FeO = 1.3$ at slag basicity of $B = 3.2$ and $PCO = 1.5 \times 101.325 kPa$ (average CO pressure in the vessel), we obtain the following equilibrium relation between the carbon content of steel and the iron oxide content of slag.

$$K = 108.8$$

$$aFeO = 1.3 N_{FeO} \approx \frac{1.3}{72 \times 1.65}(\%FeO) = 0.011 \times (\%FeO)$$

$$(\%FeO)[\%CO] = 1.25 \qquad (2.5)$$

2.1.1.4 Oxidation of Chromium

There are two valencies of chromium (Cr^{2+} and Cr^{3+}) dissolved in the slag. The ratio Cr^{2+}/Cr^{3+} increases with an increasing temperature, decreasing oxygen potential and decreasing slag basicity. Under steelmaking conditions, i.e. in the basic slags and at high oxygen potentials, the trivalent chromium predominates in the slag. The equilibrium distribution of chromium between slag and metal for basic steelmaking slags, determined by various investigators, is shown in Fig. 2.4; slope of the line represents an average of these data.

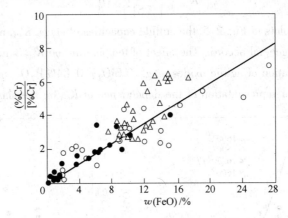

Fig. 2.4 Variation of chromium distribution ratio with the iron oxide content of slag, in the open hearth and electric arc furnace165 at tap is compared with the results of laboratory experiments

$$\frac{(\%Cr)}{[\%Cr]} = (0.3 \pm 0.1) \times (\%FeO) \qquad (2.6)$$

2.1.1.5 Oxidation of Phosphorus

It was in the late 1960s that the correct formulation of the phosphorus reaction was at last realized, thus

$$[P] + 5/2[O] + 3/2(O^{2-}) \rightleftharpoons (PO_4^{3-}) \qquad (2.7)$$

At low concentrations of [P] and [O], as in most of the experimental melts, their activity coefficients are close to unity, therefore mass concentrations can be used in formulating the equilibrium relation K_{PO} for the above reaction.

$$K_{PO} = \frac{(\%P)}{[\%P]}[\%O]^{-5/2} \qquad (2.8)$$

The equilibrium relation K_{PO}, known as the phosphate capacity of the slag, depends on temperature and slag composition.

From a reassessment of all the available experimental data, it was concluded that CaO and MgO components of the slag, had the strongest effect on the phosphate capacity of the slag. Over a wide range of slag composition and for temperatures of 1550℃ to 1700℃, the steel-slag equilibrium with respect to the phosphorus reaction may be represented by the equation

$$\lg K_{PO} = \frac{21740}{T} - 9.87 + 0.071 \times BO \qquad (2.9)$$

where $BO = \%CaO + 0.3(\%MgO)$.

2.1.1.6 Reduction of Sulfur

The sulfur transfer from metal to slag is a reduction process as represented by this equation

$$[S] + (O^{2-}) \Longrightarrow (S^{2-}) + [O] \qquad (2.10)$$

for which the state of slag-metal equilibrium is represented by

$$K_{SO} = \frac{(\%S)}{[\%S]}[\%O] \qquad (2.11)$$

As is seen from the plots in Fig. 2.5, the sulfide capacities of slags, K_{SO}, measured in three independent studies are in general accord. The effect of temperature on K_{SO} is masked by the scatter in the data. The concentration of acidic oxides, e.g. $\%SiO_2 + 0.84\%P_2O_5$, rather than the slag basicity seems to be better representation of the dependence of K_{SO} on the slag composition.

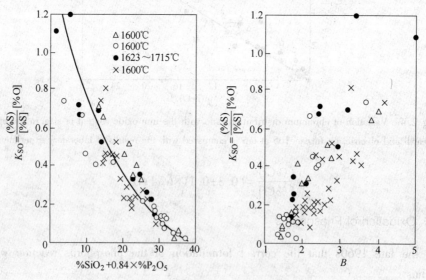

Fig. 2.5 Sulfide capacities of slags

2.1.2 State of Reactions in Steelmaking

There are numerous versions of oxygen steelmaking such as top blowing (BOF, BOP, LD, etc.), bottom blowing (OBM, Q-BOP, etc.) and combined blowing (K-OBM, LBE, etc.). The state of the refining reactions depends to some degree on the process. Bottom blowing in general provides better slag-metal mixing and reactions are closer to equilibrium. In the following section the state of reactions for top blowing (BOF) and bottom blowing (OBM) is given. The state of reactions for combined or mixed blowing processes would be between these two limiting cases. The analytical plant data used in this study were on samples taken from the vessel at first turndown from the BOP and Q-BOP shops of U. S. Steel. These plant data were acquired through the kind collaboration of the USS research personnel at the Technical Center.

2.1.2.1 Decarburization and FeO Formation

The most important reaction in steelmaking is decarburization. It not only determines the process time but also the FeO content of the slag, affecting yield and refining. When oxygen is injected into an oxygen steelmaking furnace a tremendous quantity of gas is evolved, forming a gas-metal-slag emulsion which is three to four times greater in volume than the non-emulsified slag and metal. The chemical reactions take place between the metal droplets, the slag and gas in the emulsion.

These reactions have been observed in the laboratory using x-ray techniques indicating in many cases, the gas phase (primarily CO) separates the slag and metal, and that gaseous intermediates play a role in decarburization.

When oxygen first contacts a liquid iron-carbon alloy it initially reacts with iron according to reaction (2.12), even though thermodynamically it favors its reaction with carbon. This is due to the relative abundance of iron in comparison to carbon. Carbon in the liquid metal then diffuses to the interface reducing the FeO by reaction (2.13). The net reaction is the oxidation of carbon, reaction (2.14).

$$Fe + 1/2 O_2 = FeO \tag{2.12}$$

$$FeO + C = CO + Fe \tag{2.13}$$

$$C + 1/2 O_2 = CO \tag{2.14}$$

However, carbon is only oxidized as fast as it can be transferred to the surface. At high carbon contents the rate of mass transfer is high such that most of the FeO formed is reduced and the rate of decarburization is controlled by the rate of oxygen supply:

$$\frac{d\%C}{dt} = -\frac{N_{O_2} M_C 100}{W}(f+1) \tag{2.15}$$

Where N_{O_2} ——the flow rate of oxygen in moles;

M_C ——the molecular weight of carbon (12);

W ——the weight of steel;

f ——the fraction of the product gas which is CO; the remainder is CO_2 and f is close to unity (0.8-1).

Below the critical carbon content the rate of mass transfer is insufficient to react with all the injected oxygen. In this case, the rate of decarburization is given by:

$$\frac{d\%C}{dt} = -\frac{\rho}{W}(\%C - \%C_e)\sum_i m_i A_i \tag{2.16}$$

Where %C_e——the equilibrium carbon with the slag for reaction (2.13) and is close to zero;
m_i——the mass transfer coefficient for the specific reaction site;
A_i——the metal-FeO surface area for the specific reaction site.

For top blown processes the reaction takes place between the metal droplets ejected into the emulsion and the FeO in the slag. For bottom blowing there is less of an emulsion and the reaction takes place at the interface of the metal bath and the rising bubbles which have FeO associated with them. The critical carbon is when the rates given by equation (2.15) and equation (2.16) are equal and is typically about 0.3% C.

The actual values of m_i and A_i are not known, consequently an overall decarburization constant (K_C) can be defined and the rate below the critical carbon content is given by equation (2.18) and equation (2.19):

$$K_C = \frac{\rho}{W} \sum_i m_i A_i \qquad (2.17)$$

$$\frac{d\%C}{dt} = -K_C(\%C - \%C_e) \qquad (2.18)$$

$$\ln \frac{(\%C - \%C_e)}{(\%C_C - \%C_e)} = -K_C(t - t_C) \qquad (2.19)$$

where t_C is the time at which the critical carbon content is obtained. The equilibrium carbon content (%C_e) is close to zero. However in actual steelmaking there is a practical limit of about 0.01 to 0.03 for %C_e.

The value of K_C increases with the blowing rate since the amount of ejected droplets and bubbles increases. Also, K_C decreases with the amount of steel, as indicated by equation (2.18). In actual processes, the blowing rate is proportional to the weight of steel and, therefore, K_C has similar values in most oxygen steelmaking operations. For top blowing, K_C is about 0.015s^{-1} and for bottom blowing 0.017s^{-1}. The values of the critical carbon content are also similar ranging from 0.2% to 0.4% C. The rate of decarburization for a typical steelmaking process is shown in Fig. 2.6. At the initial stage when silicon is being oxidized, the rate of decarburization is low. The value of K_C is higher for bottom blowing because mixing is more intensive and the reaction occurs at the interface of the rising bubbles as well as the slag-metal emulsion.

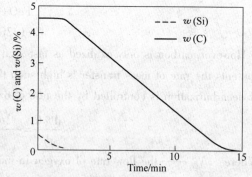

Fig. 2.6 Computed carbon and silicon contents of steel in oxygen steelmaking processes

Below the critical carbon content, when the rate of mass transfer of carbon is insufficient to reduce all of the FeO formed and, therefore, the FeO content of the slag increases rapidly. In actual processes, initially, some FeO forms because it has a low activity in the slag resulting in about 5%-

10% FeO in the slag. The FeO remains constant until the critical carbon content and then the FeO increases rapidly. The amount of FeO can be computed from a mass balance for oxygen. Specifically the oxygen not used for carbon, silicon or manganese oxidizes iron to FeO. The moles of FeO in the slag at anytime (N_{FeO}) is given by:

$$N_{FeO} = \int_0^t 2\left[\dot{N}_{O_2} - \frac{d\%C}{dt}\frac{W(1+f)}{M_C 100}\right]dt - (N_{O_2}^{Si} + N_{O_2}^{Mn}) \qquad (2.20)$$

where $N_{O_2}^{Si}$ and $N_{O_2}^{Mn}$ are the moles of oxygen consumed in oxidizing Si and Mn. Since the rate of decarburizaiton is slightly higher for bottom and mixed blowing these processes have lower FeO contents.

At all levels of turndown carbon, the iron oxide content of BOF slag is about twice that of OBM (Q-BOP) slag (Fig. 2.7). The dotted line depicts the slag-metal equilibrium value of (%FeO)[%C]≫1.25.

As depicted in Fig. 2.8, the square root correlation also applies to the product (%FeO)[%C] for low carbon contents, with marked departure form this empirical correlation at higher carbon contents. The slopes of the lines for low carbon contents are given below.

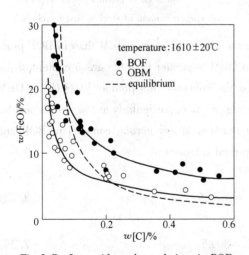

Fig. 2.7 Iron oxide-carbon relations in BOF and OBM (Q-BOP) are compared with the average equilibrium relation

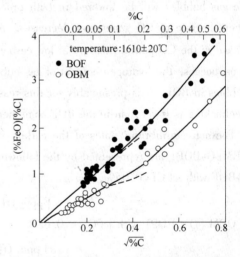

Fig. 2.8 Variation of product (%FeO)[%C] with carbon content of steel at first turndown

BOF with $w[C]<0.10\%$,

$$(\%FeO)\sqrt{\%C} = 4.2\pm0.3 \qquad (2.21)$$

OBM (Q-BOP) with $w[C]<0.10\%$,

$$(\%FeO)\sqrt{\%C} = 2.6\pm0.3 \qquad (2.22)$$

2.1.2.2 Oxygen-Carbon Relation

Noting from the plot in Fig. 2.9, there is an interesting correlation between the product [ppm O][%C] and the carbon content of the steel. The square root correlation holds up to about 0.05% C in BOF and up to 0.08% C in OBM(Q-BOP) steelmaking. At low carbon contents, the oxygen content of steel in the BOF practice is higher than that in OBM steelmaking. For carbon contents above 0.15% the product [ppm O][%C] is essentially constant at about 30±2, which is the equilibrium value for an average gas (CO) bubble pressure of about 1.5 atmosphere in the steel bath.

At low carbon levels in the melt near the end of the blow, much of the oxygen is consumed by the oxidation of iron, manganese and phosphorus, resulting in a lower volume of CO generation. With the bottom injection of argon in the BOF combined-blowing practice, and the presence of hydrogen in the gas bubbles in OBM, the partial pressure of CO in the gas bubbles will be lowered in both processes when the rate of CO generation decreases. A de-

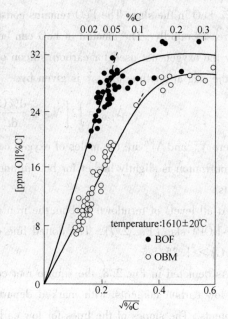

Fig. 2.9 Variation of product [ppm O][%C] with carbon content of steel at first turndown

crease in the CO partial pressure at low carbon contents will be greater in OBM than in BOF practices, because the hydrogen content of gas bubbles in OBM is greater than the argon content of gas bubbles in BOF. It is presumably for this reason that the concentration product [O][C] in OBM steelmaking is lower than in the BOF combined-blowing practice, particularly at low carbon levels.

The non-equilibrium states of the carbon-oxygen reaction at low carbon contents in BOF and OBM(Q-BOP) are represented by the following empirical relations.

BOF with $w(C) < 0.05\%$,

$$[\text{ppm O}]\sqrt{\%C} = 135 \pm 5 \tag{2.23}$$

OBM (Q-BOP) with $w(C) < 0.08\%$,

$$[\text{ppm O}]\sqrt{\%C} = 80 \pm 5 \tag{2.24}$$

2.1.2.3 Desiliconization

Silicon is oxidized out of hot metal early in the process. The oxidation reaction supplies heat and the SiO_2 reacts with the CaO to form the slag.

$$Si + 1/2 O_2 =\!=\!= (SiO_2)$$

The thermodynamics of the reaction indicate virtually all of the Si is oxidized. The rate is controlled by liquid phase mass transfer, represented by the following equation:

$$\frac{d\%Si}{dt} = -\frac{A\rho m_{Si}}{W}[\%Si - \%Si_e] \tag{2.25}$$

Where m_{Si} ——mass transfer coefficient for silicon;

$\%Si_e$ ——the silicon content in the metal in equilibrium with the slag and is close to zero.

As with decarburization, an overall rate parameter for Si can be defined, K_{Si}, similar to K_C. Within our ability to estimate K_{Si}, it is about equal to K_C since A, W, and are the same and mass transfer coefficients vary only as the diffusivity to the one half power. Therefore equation (2.25) simplifies to:

$$\ln \frac{\%Si}{\%Si°} = -K_{Si}t \qquad (2.26)$$

where $\%Si°$ is the initial silicon and K_{Si} is the overall constant for Si and is approximately equal to K_C. The rate for Si oxidation is shown in Fig. 2.6.

2.1.2.4 Manganese Oxide-Carbon Relation

At low carbon contents, the ratio $[\%Mn]/(\%MnO)$ is also found to be proportional to $\sqrt{\%C}$, represented as given below.

BOF with $w(C) < 0.10\%$,

$$\frac{[\%Mn]}{(\%MnO)} \frac{1}{\sqrt{\%C}} = 0.1 \pm 0.02 \qquad (2.27)$$

OBM (Q-BOP) with $w(C) < 0.10\%$,

$$\frac{[\%Mn]}{(\%MnO)} \frac{1}{\sqrt{\%C}} = 0.2 \pm 0.02 \qquad (2.28)$$

2.1.2.5 FeO-MnO-Mn-O Relations

From the foregoing empirical correlations for the non-equilibrium states of reactions involving the carbon content of steel, the relations obtained for the reaction of oxygen with iron and manganese are compared in the Table 2.1 with the equilibrium values for temperatures of $1610 \pm 20°C$ and slag basicities of $B = 3.2 \pm 0.6$.

Table 2.1 The relations obtained for the reaction of oxygen with iron and manganese

Composition	BOF $w(C) < 0.05\%$	OBM(Q-BOP) $w(C) < 0.08\%$	Values for slag-metal equilibrium
$\frac{[ppmO]}{(\%FeO)}$	32±4	32±5	26±9
$\frac{[\%Mn][ppmO]}{(\%MnO)}$	13.6±3.2	16.1±2.6	18±6
$\frac{(MnO)}{(\%FeO)[\%Mn]}$	2.6±0.5	2.2±0.2	1.9±0.2

It is seen that the concentration ratios of the reactants in low carbon steel describing the states of oxidation of iron and manganese, are scattered about the values for the slag-metal equilibrium. However, as indicated by the plant data in Fig. 2.10, the oxidation of iron and manganese are in the non-equilibrium states for high carbon contents in the steel at turndown. Although the concentration product $[O][C]$ is close to the equilibrium value for an average CO pressure of about $1.5\times$

101.325kPa in the steel bath, the concentrations of iron oxide and manganese oxide in the slag are above the equilibrium values for high carbon contents in the melt.

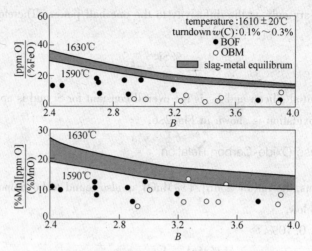

Fig. 2.10 Non-equilibrium states of oxidation of iron and manganese at high carbon contents in steel

It is concluded from these observations that:

1. At low carbon contents the equilibrium state of iron and manganese oxidation controls the concentration of dissolved oxygen.

2. At high carbon contents it is the CO-C-O equilibrium which controls the oxygen content of the steel.

2.1.2.6 State of Phosphorus Reaction

The slag/metal phosphorus distribution ratios at first turndown in BOF and OBM (Q-BOP) steelmaking are plotted in Fig. 2.11 against the carbon content of the steel. The plant data are for the turndown temperatures of $1610 \pm 20°C$ and slags containing $50\% \pm 2\%$ CaO and $6\% \pm 2\%$ MgO. The shaded area is for the slag-metal equilibrium for the above stated conditions in OBM (Q-BOP), based on the empirical [O][C] relations, i.e. for $w[C] < 0.08\%$, $[ppm\ O]\sqrt{\%C} = 80$ and for $w[C] > 0.15\%$, $[ppm\ O][\%C] = 30$.

Noting that the oxygen contents of the steel at low carbon levels are greater in BOF steelmaking, the equilibrium phosphorus distribution ratios will likewise be greater in BOF than in (OBM) Q-BOP steelmaking. For example, at $w(C) = 0.05\%$ and

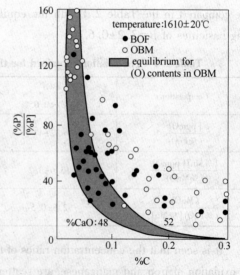

Fig. 2.11 Slag/metal phosphorus distribution ratios at first turndown in BOF and OBM (Q-BOP) practices; slag-metal equilibrium values are within the hatched area for Q-BOP

about 600ppm O in BOF at turndown the average equilibrium value of $(\%P)/[\%P]$ is about 200 at 1610℃, as compared to the average value of 60 in OBM(Q-BOP) at $w(C)=0.05\%$ with about 360ppm O.

Below 0.04% C the phosphorus distribution ratios in OBM(Q-BOP) are in general accord with the values for slag-metal equilibrium. However, at higher carbon contents the ratios $(\%P)/[\%P]$ are well above the equilibrium values. In the case of BOF steelmaking below 0.1% C at turndown, the ratios $(\%P)/[\%P]$ are much lower than the equilibrium values. On the other hand, at higher carbon contents the phosphorus distribution ratios are higher than the equilibrium values as in the case of OBM(Q-BOP).

The effect of temperature on the phosphorus distribution ratio in OBM(Q-BOP) is shown in Fig. 2.12 for melts containing 0.014% to 0.022% C with BO = 52%±2% in the slag.

2.1.2.7 State of Sulfur Reaction

A highly reducing condition that is required for extensive desulfurization of steel is opposite to the oxidizing condition necessary for steel making. However, some desulfurization is achieved during oxygen blowing for decarburization and dephosphorization. As seen from typical examples of the BOF and OBM(Q-BOP) plant data in Fig. 2.13, the state of steel desulfurization at turndown, described by the expression $[\%O](\%S)/[\%S]$, is related to the SiO_2 and P_2O_5 contents of the slag. Most of the points for OBM(Q-BOP) are within the hatched area for the slag-metal equilibrium reproduced. However, in the case of BOF steelmaking the slag/metal sulfur distribution ratios at turndown are about one-third or one-half of the slag-metal equilibrium values. At higher carbon

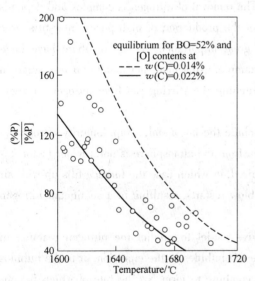

Fig. 2.12 Effect of turndown temperature on the slag/metal phosphorus distribution ratios in OBM (Q-BOP) for turndown carbon contents of 0.014 to 0.022% C; curves are for slag-metal equilibrium

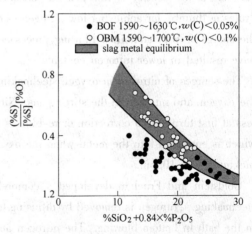

Fig. 2.13 Equilibrium and non-equilibrium states of sulphur reaction in OBM(Q-BOP) and BOF steelmaking

contents, e. g. $w(C) > 0.1\%$, the sulfur distribution ratios in both processes are below the slag-metal equilibrium values.

2.1.2.8 Hydrogen in BOF and OBM(Q-BOP) Steelmaking

Hydrogen and nitrogen are removed from the steel bath during oxygen blowing by CO carrying off H_2 and N_2.

$$2H(\text{in steel}) \longrightarrow H_2(g) \tag{2.29}$$

$$2N(\text{in steel}) \longrightarrow N_2(g) \tag{2.30}$$

There is however continuous entry of both hydrogen and nitrogen into the bath by various means. In both BOF and OBM(Q-BOP) there is invariably some leakage of water from the water cooling system of the hood into the vessel. In the case of OBM(Q-BOP), the natural gas (CH_4) used as a tuyere coolant is a major source of hydrogen.

The hydrogen content of steel in the tap ladle, measured by a probe called HYDRIS, is less than 5ppm in BOF steelmaking and 6 to 10ppm H in OBM(Q-BOP) steelmaking. Because natural gas (CH_4) is used as a tuyere coolant in the OBM, the hydrogen content of steel made in this vessel is always higher than that in the BOF vessel. In both practices the re-blow will always increase the hydrogen content of the steel. A relatively small volume of CO evolved during the re-blow cannot overcome the hydrogen pickup from various sources.

2.1.2.9 The Nitrogen Reaction

As demonstrated, the decarburization reaction is reasonably well understood as well as the slag metal re-fining reactions which approach equilibrium. The removal of nitrogen is complex and depends on numerous operating variables. With the advent of the production of high purity interstitial free (IF) steels the control of nitrogen has become of great importance. For years there have been "rules of thumb" for achieving low nitrogen. For example, using more hot metal and less scrap in the charge or with combined blowing processes, switching the stirring gas from nitrogen to argon have resulted in lower nitrogen contents.

The sources of nitrogen in oxygen steelmaking include the hot metal, scrap, impurity nitrogen in the oxygen, and nitrogen in the stirring gas. Nitrogen from the atmosphere is not a major factor unless at first turndown a correction or re-blow is required, in which case the furnace fills up with air which is entrained into the metal when the oxygen blow restarts, resulting in a significant nitrogen pickup of 5 to 10ppm.

Goldstein and Fruehan developed a comprehensive model to predict the nitrogen reaction in steelmaking. Nitrogen is removed by diffusing to the CO bubbles in the emulsion, or to the bubbles in the bath in bottom blowing. The nitrogen atoms combine to form N_2, the rate of which is controlled by chemical kinetics and the nitrogen gas is removed with the CO bubbles. Both mass transfer and chemical kinetics contribute to the overall rate of removal.

$$N(\text{metal}) = N(\text{surface})$$

$$2N(\text{surface}) = N_2 \tag{2.31}$$

The mixed control model presented previously is the basis for the model. There are, however, several complications. For example, all of the rate parameters are functions of temperature. The sulfur content of the metal changes with time and the oxygen content at the surface are considerably different from the bulk concentration. Consequently the temperature and surface contents of oxygen and sulfur must be known as functions of time. Also, scrap is melting during the process changing the composition of the metal. These details are dealt with and discussed in detail elsewhere.

An example of the results of the model calculations are shown in Fig. 2.14 and Fig. 2.15. In Fig. 2.14 the nitrogen contents are shown for two hypothetical cases for a 200 ton oxygen steelmaking converter blowing oxygen at 800Nm^3/min. In one case the charge contains 80% heavy scrap, 16 cm thick containing 50ppm nitrogen; in the second case there is no scrap but cooling is achieved by the use of a 15% addition of DRI containing 20ppm nitrogen. For the case of the heavy scrap, it melts late in the process releasing its nitrogen after most of the CO is generated, causing the nitrogen content to increase to 30ppm at the end of the blow. With no scrap the final nitrogen

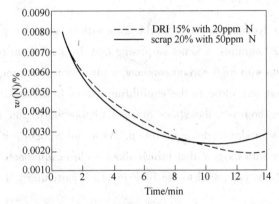

Fig. 2.14 Computed nitrogen content in a 200 ton oxygen steelmaking converter for 20% heavy scrap-80% light scrap and DRI at a blowing rate of 800Nm^3/min O$_2$

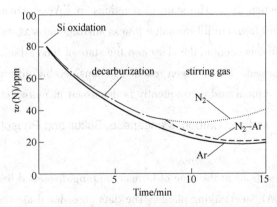

Fig. 2.15 The effect of bottom stirring practice on nitrogen removal in the BOF with N$_2$ during the entire blow, Ar during the entire blow, and with a switch from N$_2$ to Ar at 50% through the blow

is significantly lower at 20ppm. The model also indicates if ore, which has no nitrogen, is used for cooling, levels of 10ppm nitrogen can be obtained. In comparing the bulk and surface nitrogen contents, it is found that the surface concentration is only slightly less than the bulk, indicating that the nitrogen reaction is primarily controlled by chemical kinetics. In particular, towards the end of the blow the oxygen content at the surface is high, retarding the rate. This model is useful for optimizing the process for nitrogen control. For example, the effect of oxygen purity, the time of switching the stirring gas from N_2 to argon, the metal sulfur content and the amount and size of scrap can be determined. The effect of switching the stirring gas from N_2 to Ar is shown in Fig. 2.15.

2.1.2.10 General Considerations

The analyses of plant data on slag and metal samples taken at first turndown have revealed that there are indeed equilibrium and non-equilibrium states of reactions in oxygen steelmaking. In all the reactions considered, the carbon content of steel is found to have a decisive effect on the state of slag-metal reactions.

Considering the highly dynamic nature of steelmaking with oxygen blowing and the completion of the process in less than 20 minutes, it is not surprising that the slag-metal reactions are in the non-equilibrium states in heats with high carbon contents at turndown. With regard to low carbon heats all the slag-metal reactions are close to the equilibrium states in bottom blown processes. In the case of BOF steelmaking however, the states of steel dephosphorization and desulfurization are below the expected levels for slag-metal equilibrium. As we all recognize it is of course the bottom injection of lime, together with oxygen, that brings about a closer approach to the slag-metal equilibrium in OBM(Q-BOP) as compared to the BOF practice, particularly in low carbon heats.

2.2 Fundamentals of Reactions in Electric Furnace Steelmaking

The slag-metal equilibrium for EAF refining reactions are similar to those for oxygen steelmaking discussed in detail in Section 2.1. The state of reactions in EAF steelmaking are similar but, in general, are slightly further from equilibrium due to less stirring and slag-metal mixing as compared to oxygen steelmaking. In this section, the slags and the state of the carbon, sulfur and phosphorous reactions are briefly discussed. The nitrogen reaction is considerably different and, along with control of residuals, is more critical and consequently is discussed in more detail.

2.2.1 Slag Chemistry and the Carbon, Manganese, Sulfur and Phosphorus Reactions in the EAF

The state of slag-metal reactions at the time of furnace tapping discussed here is based on about fifty heats acquired from EAF steelmaking plants. The data considered are for the grades of low alloy steels containing 0.05% to 0.20% C and 0.1% to 0.3% Mn plus small amounts of alloying elements.

The iron oxide, silica and calcium oxide contents of slags are within the hatched areas shown in

Fig. 2.16. The silica contents are slightly higher than those in oxygen steelmaking slags. On the other hand the CaO contents of EAF slags are about 10% lower than those in the BOF slags. This difference is due to the higher concentrations of Al_2O_3, Cr_2O_3 and TiO_2 in the EAF slags; $w(Al_2O_3) = 4\%$-12%, $w(Cr_2O_3) = 1\%$-4% and $w(TiO_2) = 0.2\%$-1.0%. In these slags the basicity ratio B is about 2.5 at 10% FeO, increasing to about 4 at 40% FeO.

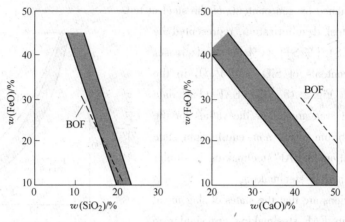

Fig. 2.16 FeO, SiO_2 and CaO contents of EAF slags at tap are compared with the BOF slags

As is seen from the plant data within the hatched area in Fig. 2.17, the iron oxide contents of EAF slags are much higher than those of the BOF slags for the same carbon content at tap. However, the product [%C][ppm O] at tap is about 26 ± 2 which is slightly lower than that for oxygen steelmaking.

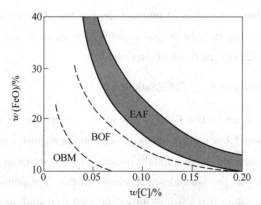

Fig. 2.17 Relation between $w(FeO)$ in slag and $w[C]$ in steel at tap in EAF steelmaking is compared with the relations in BOF and OBM(Q-BOP) steelmaking

For the slag-metal reaction involving [Mn], (FeO) and (MnO), the equilibrium relation K'_{FeMn} = (%MnO)/(%FeO)[%Mn] is 1.8±0.2 at 1625±15℃ and the basicity B>2.5. The EAF data are scattered about K'_{FeMn} = 1.8±0.4, which is near enough to the slag-metal equilibrium similar to those in oxygen steelmaking.

For steels containing 0.07% to 0.09% C in the furnace at tap, the estimated dissolved oxygen contents will be about 370 to 290ppm O.

For slags containing 38%-42% CaO and 5%-7% MgO, the equilibrium constant K_{PO} for the phosphorus reaction, is in the range 2.57×10^4 to 4.94×10^4 at 1625℃. For the tap carbon contents and slag composition ratios will be in the range 4 to 13. The EAF plant data show ratios (%P)/[%P] from 15 to 30. Similar to the behavior in BOF steelmaking, the state of phosphorus oxidation to the slag in EAF steelmaking is greater than would be anticipated from the equilibrium considerations for carbon, hence oxygen contents of the steel at tap.

The state of steel desulfurization, represented by the product $\{(\%S)/[\%S]\} \times (\%FeO)$, decreases with increasing contents of SiO_2 and P_2O_5 in the slag, as shown in Fig. 2.18. The EAF plant data within the shaded area are below the values for the slag-metal equilibrium. This non-equilibrium state of the sulfur reaction in EAF steelmaking is similar to that observed in BOF steelmaking.

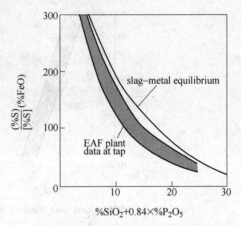

Fig. 2.18 Slag/metal sulfur distribution ratios at tap in EAF steelmaking are compared with the slag-metal equilibrium values

General indications are that the states of slag-metal reactions at tap in EAF steelmaking are similar to those noted in oxygen steelmaking at carbon contents above 0.05% C. That is, the slag/metal distribution ratio of manganese and chromium are scattered about the equilibrium values; the phosphorus and sulfur reactions being in non-equilibrium states. Also, the iron oxide content of the slag is well above the slag-metal equilibrium values for carbon contents of steel at tap. On the other hand, the product [%C][ppm O] = 26 ± 2 approximately corresponds to the C-O equilibrium value for gas bubble pressures of $(1.3 \pm 0.2) \times 101.325$ kPa in the EAF steel bath.

2.2.2 Control of Residuals in EAF Steelmaking

The growth of EAF steelmaking in the 1990's was in the production of higher grades of steel, including flat rolled and special bar quality steels. These steels require lower residuals (Cu, Ni, Sn, Mo, etc.) than merchant long products normally produced from scrap in an EAF. It is possible to use high quality scrap such as prompt industrial scrap, but this is expensive and not always available. The solution to the problem is the use of direct reduced iron such as DRI, HBI, iron carbide or other scrap substitutes such as pig iron.

The control of residuals in EAF steelmaking is based on a simple mass balance. Residual elements such as copper will report almost entirely to the steel. The control of residuals can be demonstrated in a very simple example (Table 2.2). Consider the copper content of steel produced from two grades of scrap, Number 1 bundles and shredded scrap, and DRI/HBI. It is critical to know the yield of each charge material and its copper content, which is not always easy to assess accurately. There are also secondary effects such as increased slag levels. Nevertheless, this simple example demonstrates the general concepts.

Table 2.2 Example of residuals control

Material	Yield/%	Charge/kg	Fe Yield/kg	w(Cu)/%	Cu/kg
No. 1 Bundles	95	363	345	0.05	0.17
Shredded	95	500	475	0.25	1.19
DRI/HBI	90	200	180	0.0	0
Totals		1063	1000		1.36

Note: $w(Cu)$ in the steel will be 0.136%.

2.2.3 Nitrogen Control in EAF Steelmaking

The nitrogen content of steels produced in the EAF is generally higher than in oxygen steelmaking primarily because there is considerably less CO evolution, which removes nitrogen from steel. The usual methods of controlling nitrogen are by carbon oxidation and the use of direct reduced iron or pig iron. The control of nitrogen in the EAF is discussed in recent publications by Goldstein and Fruehan. They developed a model to predict the rate of nitrogen removal similar to the one presented earlier for oxygen steelmaking. The removal of nitrogen is shown for a 100 ton EAF as a function of oxygen usage in Fig. 2.19. Nitrogen removal decreases once the carbon content falls below approximately 0.3%C, as most of the oxygen is then reacting with Fe and is therefore not producing CO. Starting at a higher initial carbon allows for more CO evolution and reduces the activity of oxygen, which retards the rate of the nitrogen reaction. The other method of reducing nitrogen is the use of DRI or HBI. These materials reduce nitrogen primarily through dilution. There was a belief that the CO evolved from these products also removed nitrogen. However, it has been demonstrated that the CO evolved from the reaction of C and FeO in the HBI/DRI is evolved at lower temperatures (1000℃) and is released while heating or in the slag phase. Since it does not pass through the metal, it does not remove nitrogen. The CO does dilute any N_2 in the furnace atmosphere.

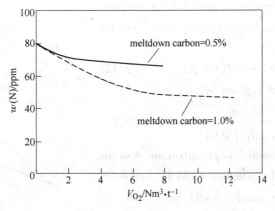

Fig. 2.19 Change in bulk nitrogen concentrations during EAF steelmaking

The nitrogen removal using 25% and 50% DRI/HBI is shown in Fig. 2.20. The primary effect is simple dilution.

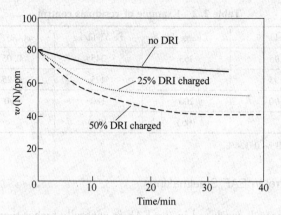

Fig. 2.20 Effect of DRI on nitrogen content during EAF steelmaking

2.3 Fundamentals of Stainless Steel Production

There are numerous processes to produce stainless steel including the AOD (argon oxygen decarburization), VOD (vacuum oxygen decarburization), and other variations of these processes. All of the processes are based on the reduction of the CO pressure to promote the oxidation of carbon in preference to chromium. This section will be limited to the reaction mechanisms and fundamentals of decarburization, nitrogen control and slag reduction.

2.3.1 Decarburization of Stainless Steel

Stainless steels can not be easily produced in an EAF or oxygen steelmaking converter as under normal conditions Cr will be readily oxidized in preference to C at low carbon contents. This leads to excessive Cr yield loss and necessitates the use of high cost low carbon ferrochrome rather than lower cost high carbon ferrochrome.

The critical carbon content at which Cr is oxidized rather than carbon can be computed based on the reaction (2.32):

$$Cr_2O_3 + 3C \rightleftharpoons 3CO + 2Cr \quad (2.32)$$

The equilibrium constant is given by:

$$K = \frac{P_{CO}^3 a_{Cr}^2}{a_{Cr_2O_3}^2 f_C^3 [\%C]^3} \quad (2.33)$$

where f_C is the activity coefficient of carbon and a_i is the activity of species i. The reaction could be written in terms of Cr_3O_4 and the results would be similar. In Fig. 2.21 the equilibrium for equation (2.33) is given as a function of P_{CO} for an 18% Cr steel.

The critical carbon content is defined as the carbon

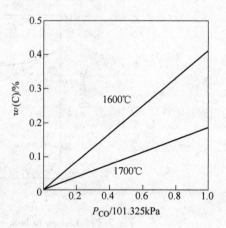

Fig. 2.21 Critical carbon content for an 18% Cr steel as a function of CO pressure

content below which Cr is oxidized. The critical carbon increases with chromium content or activity ($a_{Cr}^{2/3}$) and decreasing temperature.

However decarburization in stainless steelmaking is not controlled by equilibrium but rather reaction kinetics. It was found that for the AOD process to be effective the gas had to be injected deep in the steel bath. Fruehan demonstrated in laboratory experiments that when oxygen is injected into an Fe-Cr-C bath the oxygen initially oxidizes Cr. This lead to the following reaction mechanism and model for decarburization.

When oxygen initially contacts an Fe-Cr-C bath it primarily oxidizes Cr to Cr_2O_3. As the chrome oxide particles rise through the bath with the gas bubbles, carbon diffuses to the surface and reduces the oxide according to the equation (2.33). The rate of the reaction is controlled by mass transfer of carbon. The reaction also takes place with the top slag. The rate equation can be expressed as:

$$\frac{d\%C}{dt} = -\frac{\rho}{W}\sum_i m_i A_i [\%C - \%C_i^e] \qquad (2.34)$$

where ρ is the density of steel; W is the weight of the metal; m_i are the mass transfer coefficients; A_i are the surface areas; $\%C_i^e$ is the equilibrium carbon content; the subscript i refers to the individual reaction sites such as the rising bubble and the top slag.

The equilibrium carbon content is determined from the equilibrium for equation (2.33) which is a function of temperature, chromium content and local CO pressure. The CO pressure in turn depends on the rate of decarburization and the Ar or N_2 in the gas and the total pressure. The details of the calculation are given elsewhere.

Equation (2.35) is valid when mass transfer of carbon is limiting. At high carbon contents mass transfer of carbon is sufficient to consume all the oxygen and the rate of decarburization is simply given by the mass balance based on oxygen flow rate.

The Cr loss to the slag can be computed from the mass balance for oxygen and is given by

$$\Delta[\%Cr] = \frac{4M_{Cr}}{3W\,10^{-2}}\dot{N}_{O_2}t - \frac{10^{-2}W}{2M_C}\Delta[\%C] \qquad (2.35)$$

Where M_{Cr}, M_C——molecular weights of Cr and C;

W——weight of steel;

$\dot{N}_{O_2}$——molar flow rate of oxygen;

$\Delta[\%C]$——change in carbon content.

Simplified calculations for the rates of decarburization and Cr oxidation of an 18-8 stainless steel are presented in Fig. 2.22 and Fig. 2.23. These calculations indicate that switching the O_2/Ar ratio at 0.4% C slightly increases the rate of decarburization but significantly reduces Cr oxidation.

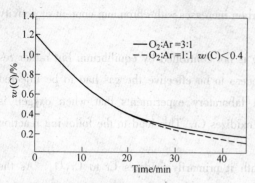 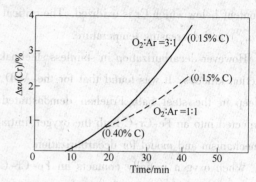

Fig. 2.22 Calculated rate of decarburization for an 18-8 stainless steel

Fig. 2.23 Calculated rate of Cr oxidation for an 18-8 type stainless steel

This model is the basis of process control models for the AOD, VOD and other similar processes.

2.3.2 Nitrogen Control in the AOD

It is desirable to use nitrogen (N_2) in the AOD in place of Ar because of its lower cost. Also, for some steels it is desirable to alloy nitrogen by blowing N_2 gas. A model for nitrogen control was developed by one of the authors based on fundamental principles and is briefly described below.

The nitrogen control model presented here is based on the mixed control model for chemical kinetics and mass transfer in series given in section 2.2.3. Equation (2.36) requires knowledge of the pressure of N_2 which is given by

$$P_{N_2} = \frac{WRTP_T}{100M_{N_2}}\frac{d\%N}{dt} + \frac{\dot{N}_{N_2}}{\dot{N}_{CO} + \dot{N}_{N_2} + \dot{N}_{Ar}} P_T \qquad (2.36)$$

Where $\dot{N}_i$ ——molar flow rate of species i;
  W——weight of steel;
  P_T——total pressure;
  M_{N_2}——molecular weight of N_2 (28).

The rate of CO evolution is determined by the rate of decarburization, equations (2.33) and (2.34), and consequently it is necessary to include the decarburization model given in the previous section.

The rate of dissociation of N_2 depends on the Cr and S contents. Glaws and Fruehan measured the rate for Fe-Cr-Ni-S alloys and found that Cr increased the rate while sulfur decreased the rate. The mass transfer parameter can be estimated from basic principles but requires knowledge of the bubble size; alternatively, it can be determined with a limited quantity of plant data by determining the best value of mA, the mass transfer coefficient times the surface area. Basic principles can be used to estimate m reasonably accurately using equations (2.37) and (2.38), thereby allowing for an estimation of A. Results of the model are presented in Fig. 2.24 and Fig. 2.25.

$$J_N = \frac{m\rho}{100}[\%N_s - \%N] \qquad (2.37)$$

2.3 Fundamentals of Stainless Steel Production

$$J_N = k\left(P_{N_2} - \frac{[\%N_s]^2}{K}\right) \quad (2.38)$$

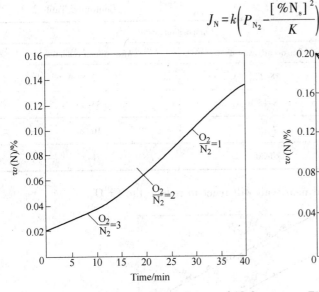

Fig. 2.24 Nitrogen pickup in 75 metric tons of 18-8 stainless steel containing 0.03% S using $O_2/N_2 = 3$ and $O_2/N_2 = 1$ gas mixtures, and with total flow of 0.85 m³/s at 1600℃

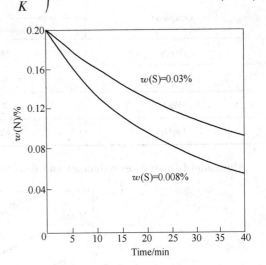

Fig. 2.25 Rate of nitrogen removal from 75 metric tons of 18-8 stainless steel containing 0.008% and 0.03% S using 0.85 m³/s of O_2 and Ar at 1600℃

2.3.3 Reduction of Cr from Slag

In stainless steel production some chromium is oxidized to slag, approximately 3% for an 18% Cr steel. It is necessary to recover this Cr by adding a reductant, usually Si, (as ferrosilicon) or Al, which reacts with Cr_2O_3. For example, for Si the reaction is given by

$$3Si + 2(Cr_2O_3) = 3(SiO_2) + 4Cr \quad (2.39)$$

The equilibrium distribution of chromium between slag and metal is given in Fig. 2.26 as a function of Si content and slag chemistry. From Fig. 2.26 and a mass balance from reaction (2.39), it is possible to compute the Si required for the desired Cr reduction. The typical compositions of AOD slags after decarburization and after silicon reduction are given in Table 2.3. As is seen from the experimental data in Fig. 2.26, the higher the slag basicity and higher the Si content of steel, the lower is the equilibrium slag/metal distribution of Cr, i.e. the greater the Cr recovery from the slag.

Table 2.3 Ranges of AOD slag composition after decarburization and after silicon reduction

Compound	Composition, w/%	
	After decarburization	After silicon reduction
FeO	4-6	1-2
MnO	4-8	1-3
SiO_2	12-18	30-40

Continued Table 2.3

Compound	Composition, w/%	
	After decarburization	After silicon reduction
Al_2O_3	18-22	3-8
CaO	8-15	33-43
MgO	7-15	10-20
Cr_2O_3	20-30	1-3

Aluminum is a stronger reductant and nearly all will react to reduce the Cr_2O_3.

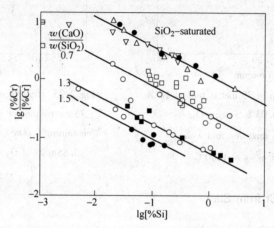

Fig. 2.26 Equilibrium slag/metal chromium distribution varying with the concentration of silicon in iron coexisting with chromium oxide containing $CaO\text{-}Al_2O_3\text{-}SiO_2$ slags at temperatures of 1600 to 1690℃

2.4 Fundamentals of Ladle Metallurgical Reactions

Several books and conferences have been devoted to ladle or secondary metallurgy. In this section only the fundamentals of deoxidation, desulfurization and inclusion modification are discussed.

2.4.1 Deoxidation Equilibrium and Kinetics

There are primarily three elements used in steel deoxidation:
1. Mn as low or high C ferro alloy.
2. Si as low or high C ferro alloy or as silico manganese alloy.
3. Al of approximately 98% purity.

2.4.1.1 Deoxidation with Fe/Mn

When the steel is partially deoxidized with Mn, the iron also participates in the reaction, forming liquid or solid Mn(Fe)O as the deoxidation product.

2.4 Fundamentals of Ladle Metallurgical Reactions

$$\left.\begin{array}{l}[Mn]+[O]\rightarrow MnO\\ [Fe]+[O]\rightarrow FeO\end{array}\right\} \text{liquid or solid Mn(Fe)O} \tag{2.40}$$

The state of equilibrium of steel with the deoxidation product Mn(Fe)O is shown in Fig. 2.27.

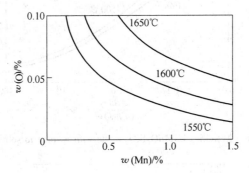

Fig. 2.27 Manganese and oxygen contents of iron in equilibrium with solid FeO-MnO deoxidation product

2.4.1.2 Deoxidation with Si/Mn

Depending on the concentrations of Si and Mn added to steel in the tap ladle, the deoxidation product will be either molten manganese silicate or solid silica.

$$\left.\begin{array}{l}[Si]+2[O]\rightarrow SiO_2\\ [Mn]+[O]\rightarrow MnO\end{array}\right\} \text{molten } xMnO\cdot SiO_2 \text{ or solid } SiO_2 \tag{2.41}$$

From the experimental work of various investigators the following equilibrium relation is obtained for the Si/Mn deoxidation reaction.

$$[Si] + 2(MnO) \rightleftharpoons 2[Mn] + (SiO)$$

$$K_{MnSi} = \left(\frac{[\%Mn]}{\alpha_{MnO}}\right)^2 \frac{\alpha_{SiO_2}}{[\%Si]} \tag{2.42}$$

$$\lg K = \frac{1510}{T} + 1.27 \tag{2.43}$$

where the oxide activities are relative to pure solid oxides. For high concentrations of silicon (> 0.4%) the activity coefficient f_{Si} should be used in the above equation, thus $\lg f_{Si} = 0.11 \times \%Si$.

The activities of MnO in manganese silicate melts have been measured by Rao and Gaskell. Their results are in substantial agreement with the results of the earlier work by Abraham, et al. The activity of the oxides (relative to solid oxides) are plotted in Fig. 2.28. For liquid steel containing $w[Mn] < 0.4\%$ the deoxidation product is a MnO-rich silicate with $w[FeO] < 8\%$; therefore the activity data in Fig. 2.28 can be used together with equation (2.42) in computing the equilibrium state of the Si/Mn deoxidation as given in Fig. 2.29. The deoxidation product being either solid silica or molten manganese silicate depends on temperature, Si and Mn contents, as shown in Fig. 2.29.

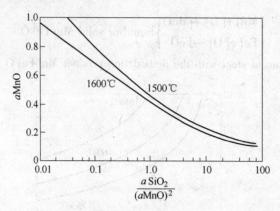

Fig. 2.28 Activities in MnO-SiO$_2$ melts with respect to solid oxides

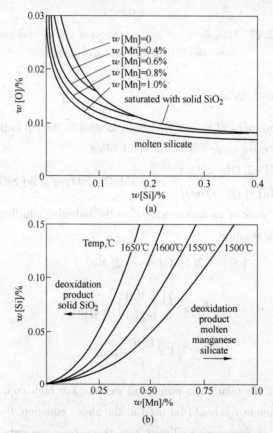

Fig. 2.29 Equilibrium relations for deoxidation of steel with silicon and manganese at 1600℃

At silica saturation, the deoxidation is by silicon alone, for which the equilibrium relation for unit SiO$_2$ activity is reduced to

$$\lg[\%Si][\text{ppm O}]^2 = -\frac{30410}{T} + 19.59 \tag{2.44}$$

2.4.1.3 Deoxidation with Si/Mn/Al

Semi-killed steels with residual dissolved oxygen in the range 40 to 25ppm are made by deoxidizing steel in the tap ladle with the addition of a small amount of aluminum together with silicomanganese, or a combination of ferrosilicon and ferromanganese. In this case, the deoxidation product is molten manganese aluminosilicate having a composition similar to $3MnO \cdot Al_2O_3 \cdot 3SiO_2$. With a small addition of aluminum, e. g. about 35kg for a 220 to 240 ton heat together with Si/Mn, almost all the aluminum is consumed in this combined deoxidation with Si and Mn. The residual dissolved aluminum in the steel will be less than 10ppm. For the deoxidation product $MnO \cdot Al_2O_3 \cdot 3SiO_2$ saturated with Al_2O_3, the silica activities are 0. 27 at 1650℃, 0. 17 at 1550℃ and decreasing probably to about 0. 12 at 1500℃. Using these activity data the deoxidation equilibria are calculated for Al/Si/Mn; these are compared in Fig. 2. 30 with the residual ppm O for the Si/Mn deoxidation at the same concentrations of Mn and Si.

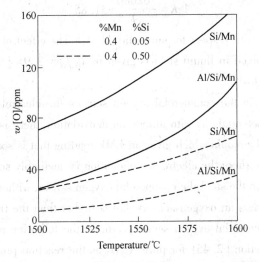

Fig. 2. 30 Deoxidation equilibria with Si/Mn compared with Al/Si/Mn for the deoxidation product saturated with Al_2O_3

When intentionally using some Al for deoxidation or unintentionally with Al being in the ferrosilicon the equilibrium deoxidation product can be solid Al_2O_3, molten manganese silicate or solid SiO_2. To avoid clogging of continuous casting nozzles, it is desirable to have liquid inclusions. The equilibrium deoxidation product is given as a function of Al and Si contents in Fig. 2. 31. As seen in this figure even small amount of Al in solution will lead to Al_2O_3 inclusions which can clog nozzles.

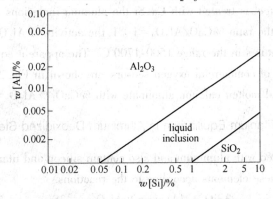

Fig. 2. 31 Equilibrium inclusions for an Fe-Al-Si-1. 0% Mn steel at 1600℃

2.4.1.4 Deoxidation with Al

Numerous laboratory experiments have been made on the aluminum deoxidation of liquid iron using the EMF technique for measuring the oxygen activity in the melt. The equilibrium constants obtained from independent experimental studies, agree within about a factor of two. An average value for the equilibrium constant is given below.

$$Al_2O_3(s) = 2[Al] + 3[O]$$

$$K = \frac{[\%Al]^2 [ppmO \times f_O]^3}{a_{Al_2O_3}} \tag{2.45}$$

$$\lg K = -\frac{62680}{T} + 31.85 \tag{2.46}$$

The alumina activity is with respect to pure solid Al_2O_3. The effect of aluminum on the activity coefficient of oxygen dissolved in liquid steel is given by $\lg f_O = 3.9 \times [\%Al]$. At low concentration of aluminum, $f_{Al} = 1.0$.

It should be noted that in the commercial oxygen sensors the electrolyte tip is MgO-stabilized zirconia. At low oxygen potentials as with aluminum deoxidation, there is some electronic conduction in the MgO-stabilized zirconia which gives an EMF reading that is somewhat higher than Y_2O_3 or ThO_2 stabilized zirconia where the electronic conduction is negligibly small. In other words, for a given concentration of Al in the steel the commercial oxygen sensor, without correction for partial electronic conduction, registers an oxygen activity that is higher than the true equilibrium value. To be consistent with the commercial oxygen sensor readings, the following apparent equilibrium constant may be used for reaction (2.45) for pure Al_2O_3 as the reaction product.

$$\lg K_a = -\frac{62680}{T} + 32.54 \tag{2.47}$$

Another point to be clarified is that in the commercial oxygen sensor system, the EMF reading of the oxygen activity is displayed on the instrument panel in terms of ppm O, as though the activity coefficient $f_O = 1.0$ in the Al-killed steel. If the deoxidized steel contains 0.05% Al, the apparent oxygen activity using equation (2.47) will be $[ppm O \times f_O] = 3.62$; noting that at 0.05% Al, $f_O = 0.64$, the apparent concentration of dissolved oxygen will be $3.62/0.64 = 5.65$ ppm O.

When the Al-killed steel is treated with Ca-Si the alumina inclusions are converted to molten calcium aluminate. For the ratio $\%CaO/Al_2O_3 = 1:1$, the activity of Al_2O_3 is 0.064 with respect to pure Al_2O_3 at temperatures in the range 1500-1700℃. The apparent equilibrium relations, consistent with the readings of commercial oxygen sensors, are shown in Fig. 2.32 for the deoxidation products: pure Al_2O_3 and molten calcium aluminate with $\%CaO/\%Al_2O_3 = 1:1$.

2.4.1.5 Silicon and Titanium Equilibrium in Aluminum Deoxidized Steel

When steels are deoxidized with aluminum and also contain silicon and titanium slag metal equilibrium is established for these elements according to the reactions.

$$3SiO_2 + 4Al = 2(Al_2O_3) + 3Si \tag{2.48}$$

2.4 Fundamentals of Ladle Metallurgical Reactions

$$3TiO_2 + 4Al \rightleftharpoons 2(Al_2O_3) + 4Ti \qquad (2.49)$$

In principle the equilibrium distribution ratio %Si/(%SiO$_2$) and %Ti/(%TiO$_2$) can be computed from basic thermodynamics. The equilibrium ratios observed in practice are shown in Fig. 2.33 and 2.34.

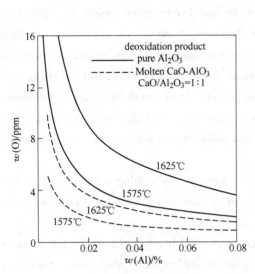

Fig. 2.32 Deoxidation with aluminum in equilibrium with Al$_2$O$_3$ or molten calcium aluminate with CaO/Al$_2$O$_3$ = 1 : 1

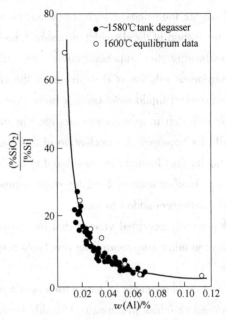

Fig. 2.33 Tank degasser data are compared with the experimental equilibrium data for aluminum reduction of silica from lime-saturated calcium aluminate melts containing $w(SiO_2) < 5\%$

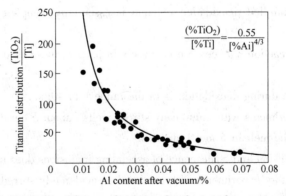

Fig. 2.34 Silicon and titanium distribution after vacuum

2.4.1.6 Rate Phenomena in Deoxidation

The equilibrium states for the common deoxidation reactions pertaining to steelmaking conditions have been established reasonably well. However, the rate phenomena concerning the deoxidation reactions is complex and is discussed in detail in a review paper by Turkdogan. A brief mention of the state of our knowledge is adequate for the present purpose.

There are three basic consecutive steps involved in the deoxidation reaction; namely, formation of critical nuclei of the deoxidation product in a homogeneous medium, progress of deoxidation resulting in growth of the reaction products and their flotation from the melt.

Turkdogan has shown that because of the high interfacial tension between liquid iron and oxide and silicate inclusions, a high supersaturation of the reactants in the metal is needed for spontaneous nucleation of the deoxidation products as predicted from the theory of homogeneous nucleation. In estimating the supersaturation ratio likely to be achieved under practical conditions, homogeneous solution of deoxidizers in the steel was assumed. However, the dissolution of added deoxidizers in liquid steel takes a finite time during which certain regions of the melt are expected to be very rich in solute concentration; in these regions the solution is sufficiently supersaturated locally for homogeneous nucleation of the deoxidation product. Owing to the agitation in the ladle, the nuclei thus formed are considered to be distributed in the melt, soon after the addition of deoxidizers. Another source of nuclei is, of course, the thin oxide layer on the surface of particles of solid deoxidizers added to steel.

A generally accepted view is that the deoxidation reactions at steelmaking temperatures are fast relative to other rate-controlling processes responsible for the growth and ultimate flotation of inclusions.

The rate phenomena is deoxidation is complex because of the side effects caused by the interplay of several variables which cannot readily be accounted for in mathematical simulations of the deoxidation process. However, certain important deductions can be made from the results of several conceptual analyses based on simplified models and those of experimental observations.

1. The number of nuclei (z) formed at the time of addition of deoxidizers is of the order of $z = 1 \times 10^7 / cm^3$ or higher.

2. The diffusion-controlled deoxidation reaction is essentially complete within a few seconds when $z > 1 \times 10^6 / cm^3$.

3. The deoxidation reaction may cease prematurely in parts of the melt depleted of nuclei or oxide inclusions.

4. The inclusion size during deoxidation is in the range 1 to 40mm.

5. In laboratory experiments with inductively stirred melts (about 5 cm deep) most of the oxide inclusions float out of the melt in 5 to 10 minutes.

6. The growth by collision and coalescence of ascending inclusions does not seem possible under the conditions of laboratory experiments with unstirred or moderately stirred melts.

These observations are not mutually consistent. One possible explanation perhaps is that the nuclei formed at the time of dissolution of deoxidizers are unevenly distributed in molten steel. In parts of the melt where the number of nuclei is small, e. g. 1×10^4 to $1 \times 10^5 / cm^3$, the inclusions 20 to 40mm in size rapidly float out of the melt, presumably prior to the deoxidation reaction. In parts of the melt containing about 10^8 nuclei/cm^3, the inclusions grow only to a micron size and ascend in the melt with a creeping velocity. Convection currents or other means of stirring eventually bring about more uniform distribution of these small inclusions. The particles thus brought to the parts of the melt where the deoxidation reaction was incomplete, due to early depletion of inclusions, bring about further deoxidation, growth and flotation.

Under practical conditions of deoxidation during filling of the ladle, there is sufficient stirring that some inclusion growth may take place by collision and coalescence; also stirring brings about a motion in the melt such that the inclusions could get attached to the surface of the ladle lining and caught by the slag layer. Controlled gas stirring at low flow rates for 4-8 minutes is common practice to enhance inclusion removal.

2.4.2 Ladle Desulfurization

It is possible to desulfurize aluminum killed steels in the ladle or ladle furnace, using CaO based slags, to less than 20ppm S. The chemical reaction can be written as

$$3(CaO) + 2Al + 3S \rightleftharpoons 3(CaS) + Al_2O_3 \tag{2.50}$$

The equilibrium sulfur distribution ratio can be calculated from the thermodynamics of reaction (2.50). In terms of the ionic reaction the reaction is

$$2/3Al + S + (O^{2-}) \rightleftharpoons (S^{2-}) + 1/3(Al_2O_3) \tag{2.51}$$

for which the equilibrium ratio is

$$K_{SA} = \frac{(\%S)}{[\%S]}[\%Al]^{-2/3} \tag{2.52}$$

The value of K_{SA} for CaO-Al_2O_3 and CaO-Al_2O_3-SiO_2-MgO slags is shown in Fig. 2.35. The sulfur distribution ratios are shown in Fig. 2.36 and Fig. 2.37 for the stated Al contents. From the data presented and a mass balance for sulfur it is possible to compute the final equilibrium sulfur contents.

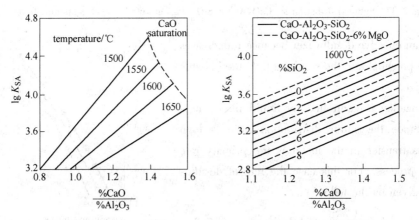

Fig. 2.35 Effects of temperature and slag composition on the equilibrium relation K_{SA} for the calcium-magnesium aluminosilicate melts

The slags will absorb sulfur until CaS forms decreasing the amount of dissolved CaO and decreasing K_{SA}. The solubility of CaS is given in Fig. 2.38.

Steels deoxidized with Si are difficult to desulfurize because the oxygen potential is significantly higher than for Al killed steels. For these steels desulfurization is limited to 10% to 20%. Calcium

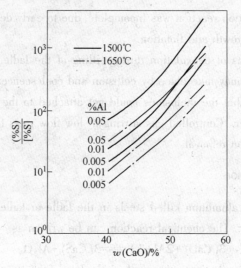

Fig. 2.36 Sulfur distribution for CaO-Al$_2$O$_3$ slags at 1600℃

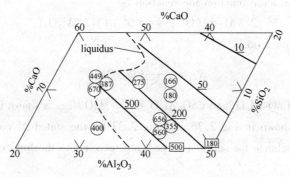

Fig. 2.37 Sulfur distribution for CaO-Al$_2$O$_3$-SiO$_2$ slags at 1600℃ steel containing 0.03% Al

carbide is an effective desulfurizer because it decreases the oxygen potential and the resulting CaO then is able to desulfurize.

Desulfurization is controlled by liquid phase mass transfer. Since the sulfur distribution ratio is high sulfur mass transfer in the metal is the primary rate controlling process. For this case the rate of desulfurization is given by the following.

$$K_{SA} = \frac{(\%S)}{[\%S]}[\%Al]^{-2/3} \qquad (2.53)$$

$$[\%S^e] = \frac{(\%S_t)}{L_S} \qquad (2.54)$$

$$(\%S_t) = \frac{W_m}{W_s}([\%S_o] - [\%S_1]) \qquad (2.55)$$

Where   m——mass transfer coefficient;

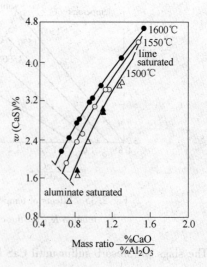

Fig. 2.38 Solubility of CaS in calcium aluminate melts related to the mass ratio %CaO/%Al$_2$O$_3$

W_m——weight of metal;

W_s——weight of steel;

r——density of steel;

L_S——sulfur distribution ratio;

$(\%S_t)$——sulfur content of the slag;

$[\%S_t]$——sulfur content of the metal;

$[\%S^e]$——equilibrium sulfur content of the metal;

$[\%S_0]$——initial sulfur content of the metal.

Equations (2.53) through (2.55) can be solved and the rate is given by

$$\frac{\ln\left\{1+\frac{1}{L_S}\left(\frac{W_m}{W_s}\right)\left[\frac{\%S_t}{\%S_0}\right]-\frac{1}{L_S}\left(\frac{W_m}{W_s}\right)\right\}}{1+\frac{1}{L_S}\left(\frac{W_m}{W_s}\right)}=\frac{mA\rho}{W_s}t \qquad (2.56)$$

Desulfurization increases with stirring rate, which increases m and more importantly A by providing slag-metal mixing, increased values of L_s, and higher slag volumes. It has been shown that 90% desulfurization can be achieved in 10-15 minutes of intense stirring.

2.4.3 Calcium Treatment of Steel

Calcium is usually employed in ladle metallurgy as Ca or CaSi cored wire or by injecting CaSi powder. Calcium is highly reactive and it could deoxidize, desulfurize, modify oxide inclusions or modify sulfide inclusions. It should only be used in deoxidized steels because it is too expensive to be used as a deoxidizer.

Since often the primary purpose of a calcium injection into the steel bath is to convert solid Al_2O_3 inclusions to liquid calcium aluminates to prevent Al_2O_3 from clogging casting nozzles, it is necessary to know under what conditions it will react with the inclusions or simply react with sulfur. There have been numerous thermodynamic calculations to predict the conditions for Al_2O_3 inclusion modification. Several of these required knowledge of the thermodynamics of Ca in steel, which is not accurately known. The conditions can be computed based on the thermodynamic properties of the inclusions themselves.

Consider the following reaction equilibrium in the calcium treated steel,

$$3(CaS) + (Al_2O_3) = 3(CaO) + 2Al + 3S \qquad (2.57)$$

where CaS is that formed by the reaction of calcium and sulfur; (Al_2O_3) is an alumina rich inclusion such as Al_2O_3, $CaO \cdot Al_2O_3$, etc; (CaO) represents an inclusion richer in CaO. The results of these calculations are given in Fig. 2.39 in which the inclusion stability is given as a function of Al and S contents. Below the CA curve $CaO \cdot Al_2O_3$ is the stabler oxide and below the $C_{12}A_7$ liquid curve, liquid calcium aluminates form. For example if Ca is added to a steel containing 0.04% Al and 0.015% S, the alumina inclusion will be converted to solid $CaO \cdot Al_2O_3$ and calcium there will react to form CaS. These inclusions will clog casting nozzles. For effective inclusion modification the sulfur content should be below 0.01% for a 0.04% Al steel.

Calcium also helps eliminate MnS inclusions which form during solidification. For a steel low in sulfur which contains liquid calcium aluminate inclusions, much of the remaining sulfur will be absorbed during cooling and solidification by these inclusions as they are excellent desulfurizers. Also, if any dissolved Ca is remaining, although the authors believe this will be very small, it could react with sulfur on the inclusions. The net result is a duplex inclusion consisting of a liquid calcium aluminate with a CaS (MnS) rim. This is preferable to stringer MnS which can cause brittle fracture.

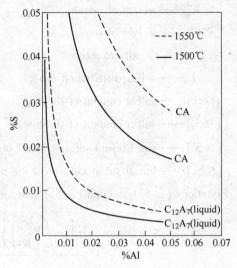

Fig. 2.39 Univariant equilibrium for CaS (aCaS = 0.74) and $C_{12}A_7$ or CA as a function of %Al and %S at 1550℃ and 1500℃

2.5 Fundamentals of Degassing

During the past two decades there has been an increase in vacuum degassing to reduce hydrogen and nitrogen contents. However, the major increase in vacuum degassing is for decarburization to low carbon contents (40ppm) which are required for good formability such as for interstitial free (IF) steels. Briefly there are two major types of degassers: circulating, such as RH and RH-OB, and non-recirculating such as ladle and tank degassers. In some cases oxygen is used to enhance the reactions, examples of these are RH-OB (oxygen blowing) and VOD (vacuum oxygen decarburization).

In this section only the fundamental aspects of the reactions will be discussed.

2.5.1 Fundamental Thermodynamics

In vacuum degassing hydrogen, nitrogen, carbon, and oxygen can be removed by the following reactions.

$$H = 1/2 H_2 \tag{2.58}$$

$$N = 1/2 N_2 \tag{2.59}$$

$$C + O(FeO, 1/2 O_2) = CO + (Fe) \tag{2.60}$$

Hydrogen and nitrogen are dissolved in steel and removed by forming diatomic molecules.

Carbon is removed by reaction with oxygen as dissolved oxygen, FeO in slag or gaseous oxygen (O_2) to form CO. Due to the reduced pressure the formation of CO is favored. This is shown schematically in Fig. 2.40 as a plot of the equilibrium carbon and oxygen contents at 1 and 0.2 atmosphere pressure. In RH and other processes with oxygen the carbon reacts with oxygen and follows the indicated reaction path. Theoretically one atom of oxygen requires one atom of dissolved oxygen. However in actual processes there are other sources of oxygen such as leaks and unstable oxides in the slag (FeO and MnO). In oxygen assisted processes the reaction path is indicated as RH-OB (or VOD). Initially gaseous oxygen removes carbon, the oxygen blow is terminated, and dissolved oxygen reacts with carbon.

2.5.2 Vacuum Degassing Kinetics

In recirculating systems H_2, N_2 and CO are formed at two reaction sites: the rising argon bubbles and the metal-vacuum interface above the melt. The model and plant data given by Bannenberg et al indicates

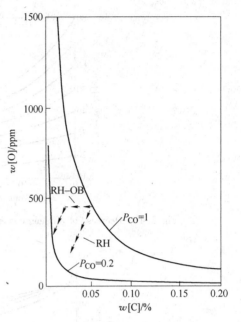

Fig. 2.40 Carbon-oxygen equilibrium as a function of CO pressure

that the Ar bubbles are the major reaction surface area for reactions at high Ar rates. Recent work by Uljohn and Fruehan indicate at lower Ar flow rates both sites should be considered. As the argon bubbles burst at the surface they generate a large number of metal droplets and surface area. In recirculating systems the reactions include the rising argon bubbles, the metal droplets and homogenous or heterogeneous nucleated bubbles in the bath.

The hydrogen reaction is simply controlled by liquid phase mass transfer of hydrogen. Hydrogen diffusivity is high and consequently the reaction is fast. For non recirculating systems the reactions increase with argon stirring in a complex manner.

It should be noted that the reaction is limited not by H_2 gas in the vessel but rather H_2O. Even at very low H_2O pressures the equilibrium hydrogen content for reaction (2.61) is relatively high.

$$3H_2O + 2Al = Al_2O_3 + 6H \qquad (2.61)$$

For example, for $P_{H_2O} = 2 \times 10^{-6}$ (101.325kPa) and an aluminum content of 0.03%, the equilibrium hydrogen content is approximately 1ppm.

The nitrogen reaction is controlled by mass transfer of nitrogen and chemical kinetics in series and the mixed control model given in Section 2.2 can be applied. Again the reactions are complex and the details are given elsewhere. However, the rate depends on the argon flow rate and sulfur content; sulfur is surface active and retards the chemical reaction as discussed in Section 2.2. Examples of calculated rates of nitrogen are given in Fig. 2.41 and Fig. 2.42. Nitrogen can only be removed effectively at low sulfur contents and high Ar bubbling rates.

The rates in recirculating systems (RH) are complicated by the fact they also depend on the cir-

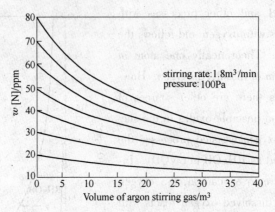

Fig. 2.41 Decrease of nitrogen content during vacuum for a sulfur content of 10×10^{-6}

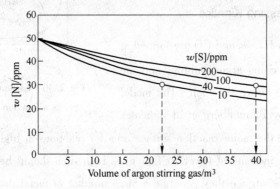

Fig. 2.42 Influence of sulfur content on the denitrogenisation

culation rate into the RH unit. For example the rate of decarburization in the ladle is given by

$$\frac{d\%C}{dt} = -\frac{Q}{W}[\%C - \%C_{RH}] \qquad (2.62)$$

where Q ——metal recirculation rate;

W ——the weight of steel;

$\%C_{RH}$ ——the carbon content in the RH.

The change of carbon in the RH is given by

$$\frac{d\%C_{RH}}{dt} = -k[\%C_{RH} - \%C_e] + \frac{Q}{W_{RH}}[\%C - \%C_{RH}] \qquad (2.63)$$

where k is the decarburization rate constant; $\%C_e$ is the equilibrium carbon content; W_{RH} is the weight of steel in the RH. An approximate solution to equation (2.62) is given by

$$\ln\left(\frac{\%C - \%C_e}{\%C_\circ - \%C_e}\right) = -Kt \qquad (2.64)$$

Where $\%C_\circ$ ——the initial carbon content;

K ——the overall reaction rate constant given by

$$K = \frac{Q}{W}\left(\frac{kW_{RH}}{kW_{RH} + Q}\right) \qquad (2.65)$$

Therefore the rate can be increased by increasing Q, W_{RH} and k. For low circulation rates the rate is controlled by the circulation rates. Consequently newer RH units have large snorkels and high circulation rates.

The hydrogen reaction in RH units are given by similar expressions. The major difference is that k_{RH} for hydrogen is larger because of faster mass transfer of hydrogen. Nitrogen removal in RH units is generally slow and generally less than 5ppm is removed.

Exercises

2-1 Please explain the mechanism of the oxidation for phosphorus.
2-2 How many reactions in EAF, and how to control the residuals?
2-3 What are the keys for stainless steel production?
2-4 What are the main tasks for the ladle metallurgical?
2-5 Please explain the mechanism of degassing process for thermodynamics and kinetics.

3 Pre-treatment of Hot Metal

3.1 Introduction

Pre-treatment of hot metal is the adjustment of the composition and temperature of blast furnace produced hot metal for optimal operation of the oxygen converter process; as such, it is one of the interdependent chains of processes that constitute modern steelmaking, as shown in Fig. 3.1. When taken to the extreme case, the converter process function is reduced to scrap melting and carbon reduction subsequent to the prior removal of silicon, phosphorus and sulfur in preparatory steps under thermodynamically favorable conditions. An important benefit of removing phosphorus and sulfur from the hot metal prior to the oxygen converter process is the ability to produce steels with phosphorus and sulfur contents lower than otherwise achievable without severe penalty to the converter process. Silicon removal is beneficial to the converter to reduce the chemical attack on the basic refractory lining and to allow the use of only minimal amounts of slag-making fluxes, thereby maximizing process yield.

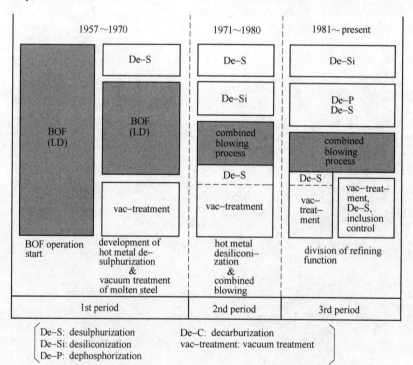

Fig. 3.1 Changes in refining functions in the Japanese steel industry

Hot metal pre-treatment by North American and European steel producers presently is focused on desulfurization due to the common use of relatively low phosphorus containing iron ores.

In a unique approach to pre-treatment, ISCOR, in South Africa, installed a hot metal mixer equipped with channel inductors to provide electrical energy to heat the liquid, and thereby raise the scrap melting capability of the steel plant.

Details of the process steps introduced above are provided in the following sections.

3.2 Desiliconization and Dephosphorization Technologies

The introduction of oxygen converter technology in Japan occurred at a time of limited availability of high quality scrap, and, as a result, the desire was to minimize the use of this expensive resource. Steel production was focused on the use of controlled, prepared raw materials. The technologies developed for the efficient removal of silicon and phosphorus from the hot metal, both fundamentally endothermic when carried out using the customary oxide reagents, provided an economic benefit by consuming chemical energy otherwise available for melting scrap in the converter. By 1983, a large number of pre-treatment facilities were in use (Table 3.1).

Table 3.1 Hot metal pre-treatment facilities in Japan (1983)

Items	Desiliconization equipment				Dephosphorization equipment				
	Desiliconization in blast furnace runner		Desiliconization transport vessel		Transport vessel (soda ash)		Transport vessel (lime-based flux)		Furnace for exclusive use (converter)
In operation	NSC	Kimitsu (No. 2, 4 BF) Yawata (No. 4 BF)	NSC	Muroran Yawata Sakai Nagoya	NSC	Yawata (No. 1 LD plant)	NSC	Kimitsu Nagoya	KSC	Chiba (No. 2 LD plant)
	NKK	Fukuyama (No. 4 BF)	NKK	(No. 1 LD plant) Fukuyama	NKK	Fukuyama		(No. 1 LD plant) Muroran Yawata		Mizushima Kobe
	KSC	Chiba (No. 6 BF)	SMI	Kashima	SMI	Kashima Kashima	KSC	Chiba (No. 1 LD plant)		
	SMI	Kashima (No. 3 BF) Wakayama (No. 4 BF)								
	KSL	Kobe (No. 3 BF)								

Continued Table 3.1

Items	Desiliconization equipment				Dephosphorization equipment			
	Desiliconization in blast furnace runner		Desiliconization transport vessel		Transport vessel (soda ash)	Transport vessel (lime-based flux)		Furnace for exclusive use (converter)
Planned or under construction	NSC	Oita	NSC	Nagoya	NISSHIN	NSC	Oita	
		(No. 2 BF)			Kure		Nagoya	
	NKK	Fukuyama		(No. 2 LD plant)			(No. 2 LD plant)	
		(No. 2 BF)	NKK	Keihin		NKK	Keihin	
		Keihin	SMI	Wakayama		KSC	Chiba	
		(No. 1 BF)	KSL	Kakogawa			(No. 3 LD plant)	
	SMI	Kokura				SMI	Wakayama	
		(No. 2 BF)					Kokura	
	NISSHIN					KSL	Kakogawa	
		Kure						
		(No. 2 BF)						

Initially, these pre-treatment processes were performed by adding iron ores or sinter to the hot metal during its flow in the blast furnace runner. Further improvements and control over chemical results were attained by the addition via subsurface injection of the reagents in dedicated vessels, such as oversized torpedo or submarine cars. This brought on the use of a variety of chemical reagents, including soda ash (sodium carbonate), which also provides for significant removal of sulfur. When using iron oxides for desiliconization, it is essential to separate, i.e., remove, the process slag before the hot metal is desulfurized as this operation requires low oxygen potential for efficient performance. It is important to recognize that phosphorous removal occurs only in hot metal containing less than 0.15% Si, additionally, phosphorus held in the slag could be subject to reduction, i.e., reversal, into the hot metal if it were present during desulfurization. An interesting technical development was the combination of dephosphorization and desulfurization in a single vessel whereby phosphorous is reacted with the oxidizing reagents as they rise in the liquid and sulfur is removed by the top slag in the vessel (Fig. 3.2).

Desiliconization and dephosphorization are accompanied by losses of carbon from the hot metal and evolution of CO_2 from carbonate reagents. Thus, control strategies such as addition of coke breeze or equipment accommodations must be made in the reaction vessel and gas capture systems to contain foaming and flame evolution. In the recent timeframe, environmental considerations over disposal of sodium-containing slag has forced the use of limestone based reagents, often mixed with

3.2 Desiliconization and Dephosphorization Technologies

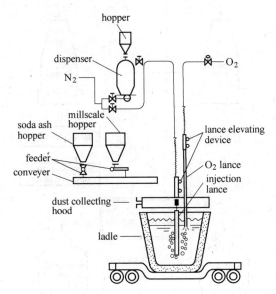

Fig. 3.2 Equipment for concurrent dephosphorization and desulfurization

iron ore or sinter fines and delivered with oxygen, the latter used to diminish the thermal penalty from the pre-treatment process. Oxygen consumption in these process steps is illustrated in Fig. 3.3.

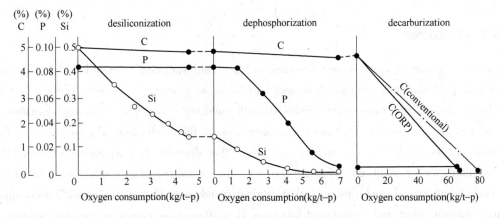

Fig. 3.3 Changes in silicon, phosphorus and carbon contents of iron in each stage of steelmaking

In some plants, the silicon and phosphorus removal steps occur in full size oxygen converter vessels and the resulting carbon containing liquid is transferred, after separation of the low basicity primary process slag, into a second converter (Fig. 3.4), for carbon removal by oxygen top blowing. In this sequence, the slag from the second vessel is used as a starter slag for the first step. In a way, this is equivalent of the former open hearth process, which provided for flushing of the initial silica and phosphorus rich slag and thus allowed the use of hot metal made from phosphorus bearing ores for production of what was then considered low phosphorus steels.

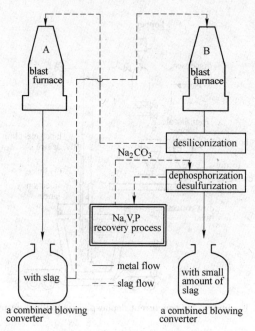

Fig. 3.4 Refining process based on soda ash treatment

3.3 Desulfurization Technology

3.3.1 Introduction

"...sulfur is frequently found in metallic ores, and, generally speaking, is more harmful to the metals, except gold, than other things. It is most harmful of all to iron...", so wrote Agricola four and one half centuries ago. From ancient times, through puddling furnaces and into blast furnaces, the control and elimination of sulfur has been a major task for the steelmaker. The cost of sulfur is enormous. In its simplest form, a modern coal to steel flow sheet involves separating more than 99% of the sulfur dug out of the ground at the coal pits. To control the 11kg/t of sulfur contributed by the coal and other feed stocks, the typical steel plant spends over $5/t of steel in addition to capital charges for equipment and exclusive of processes for sulfide shape control in the steel product.

In the oxygen driven converter operation, the heavily oxidizing environment of the metal and slag and the inability to attain the equilibrium sulfur partition ratio between slag and metal, limit the sulfur removal capability of the process. Thus, to bring the sulfur content of the steel to within the range manageable by the far more costly steel desulfurization, the lower cost hot metal treatment technologies have been developed to remove sulfur prior to the oxygen steelmaking step.

Initially these technologies were used to help the steelmaker, but, in time, it was recognized that significant cost savings and production increases in ironmaking would result if sulfur limits formerly imposed on the blast furnace operation were lifted. In most North American steel plants, the hot metal leaves the blast furnace containing 0.040%-0.070% S, while the oxygen converters are charged with hot metal containing as little as 0.010%-0.001% S, to conform to limits on steel composition set by caster operations and final product quality requirements.

The importance of sulfur management and the huge costs involved have led to worldwide efforts to develop and implement an array of different desulfurization technologies. The different reagent and delivery systems in use are the result of local economic and environmental factors and the preferences of technical and operating management at the individual plant sites.

The following sections address the main chemical reactions for sulfur removal from hot metal, the range of process permutations, the specifics of reagent delivery systems, the importance of reaction vessel selection and of slag management issues as these topics bear on a well-functioning system.

3.3.2 Process Chemistry

The variety of process permutations adopted worldwide depend on one or a combination of the following reactions:

$$Na_2CO_3(s) + S + C = Na_2S(l) + CO_2(g) + CO(g) \tag{3.1}$$
$$Mg(s) + S = MgS(s) \tag{3.2}$$
$$CaC_2 + S = CaS(s) + 2C \tag{3.3}$$
$$CaO + S + C = CaS(s) + CO(g) \tag{3.4}$$
$$Mg + CaO + S = CaS(s) + MgO(s) \tag{3.5}$$
$$CaO + 2Al + S + 3O = (CaO \cdot Al_2O_3)(S) \tag{3.6}$$
$$(CaO \cdot Al_2O_3)(s) + S = (CaO \cdot Al_2O_3)(S) \tag{3.7}$$

Initially, most plants relied on reaction (3.1), that is, the addition of soda ash (Na_2CO_3) at the blast furnace cast house or at the steelworks while filling iron transfer ladles. This approach was abandoned as process control and environmental management were very difficult. In Europe and Japan, mechanical stirrers (KR) were introduced into the blast furnace runners (Fig. 3.5), and later for use in hot metal transfer ladles (Fig. 3.6). In the U.S. and Canada, the next step was dependent on reaction (3.2) with the use of Magcoke (a product made by filling the pores of coke with magnesium and submerging this material into the hot metal in a sequence of multiple dunks). Results were reproducible, but, attainment of sulfur contents of less than 0.020% was costly in reagents and process time. Capture of the copious magnesium fumes was nearly impossible without total building evacuation.

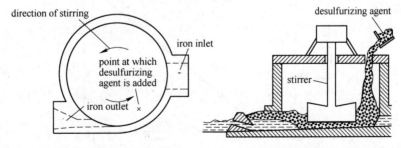

Fig. 3.5 Plan (left) and elevation (right) views of a unit for continuous desulfurization of iron

The chemical behavior of magnesium in hot metal has been the subject of extensive study

(Fig. 3.7). It is important to realize that the solubility product of Mg and S is strongly dependent on temperature (Fig. 3.8) and silicon and carbon content of the iron. This results in improved sulfur removal, or the reduced need for magnesium for colder iron. The practical effect is to lessen the cost penalty of having to load relatively cold hot metal with magnesium for attainment of sulfur levels lower than 0.002%S. An interesting technical side effect of the increase in solubility of magnesium in hot metal at low sulfur levels, e.g. near 10ppm S, is the observation that some magnesium appears to be oxidized from the iron as soon as the raker blade clears the surface of slag. The effect is noticeable as a light white fume even after emptying the transfer ladle into the steelmaking vessel. Sampling has shown the plume material to consist mostly of MgO.

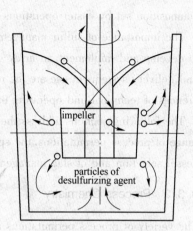

Fig. 3.6 Schematic diagram of the KR desulfurization method

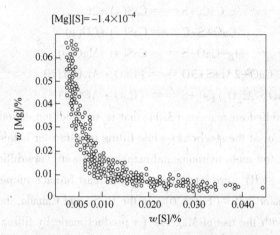

Fig. 3.7 Experimental data on magnesium and sulfur content in hot metal treated with granulated magnesium at 1400℃

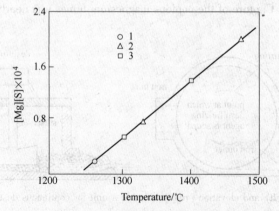

Fig. 3.8 Dependence of Mg and S product on temperature

3.3 Desulfurization Technology

A critical step in process development was adoption of subsurface pneumatic injection of calcium carbide powder [reaction (3.3)] and of pulverized lime [reaction (3.4)] or combinations, i. e., mixtures, of pulverized magnesium and lime [reaction (3.5)]. Because calcium carbide is inert, it is difficult to distribute it throughout the liquid; to improve on this, one of two reagents is added (about 15%-20%) to create surface and stirring: limestone, which cools the liquid or diamid lime, which is less endothermic. The latter version, known as CaD, was developed by SKW in Germany.

An important improvement came in the development of co-injection technology: the controlled mixing in the transport line of reagents supplied separately (Fig. 3.9). This technique, now in universal use, allows for a wide array of reagent combinations and permits independent adjustment of the rates of the delivery of the reagents during the process (Fig. 3.10). This is most useful for magnesium based systems wherein splash and fuming during lance insertion and removal can be kept to a minimum by starting and stopping magnesium reagent flow with the lance tip at the deepest immersion in the hot metal. Another advantage is that the rate of delivery of the magnesium can be reduced at low sulfur levels when low magnesium solubility limits its dissolution in the iron (Fig. 3.11). Co-injection affords a cost benefit by allowing the user to purchase the individual components from the least costly material supplier rather than being solely reliant on a supplier for a proprietary mixture. A further benefit over blended reagent mixtures is elimination of segregation (i. e., separation) of individual reagent components with differing size or density while the mixture is in transit or in storage.

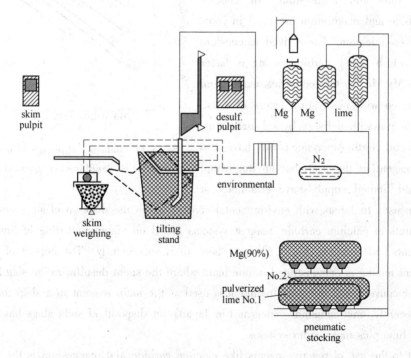

Fig. 3.9 Schematic of a hot metal treatment station using co-injection technology at LTV Steel Indiana Harbor Works

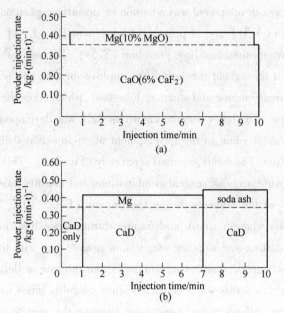

Fig. 3.10 Reagent injection patterns by co-injection

Although in Europe and Canada carbide based systems had been favored for a number of years, several European shops also have adopted co-injection for lime and magnesium. In another variant, carbide and magnesium are used in combination by co-injection. For a fixed magnesium feed rate (splash limit) carbide + Mg is faster than lime + Mg. In the CIS countries, magnesium granules, coated with passivating layers of salts, have been in use with delivery by subsurface injection. Several North American plants have also used this reagent in the past, but environmental concerns and limited supply have eliminated it

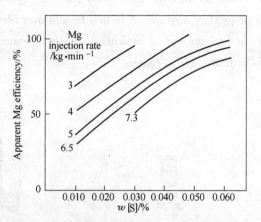

Fig. 3.11 Influence of the rate of injection of magnesium on desulfurization apparent efficiency

from current use. In Japan, with environmental constraints on the disposal of slags containing residual amounts of calcium carbide, reagent systems based on vigorous stirring of lime and soda based reagents (KR process, Fig. 3.6) have been used successfully. The benefit of intense hot metal reagent mixing is demonstrated in one plant where the spent desulfurization slag from a shop equipped for conventional injection treatment is used as the main reagent in a shop using the KR process. Recently, environmental concerns (in Japan) on disposal of soda slags has brought on adoption of lime plus magnesium systems.

A variant to the use of reactive agents like calcium carbide and magnesium is the use of lime powder either preceded by addition of aluminum to the hot metal [reaction (3.6)], or lime delivered with organic stirring agents such as natural gas and/or solid hydrocarbons. The latter, used as ground solid hydrocarbons, has been adopted to improve mixing even for magnesium based reagents. With the use of aluminum, $CaO \cdot Al_2O_3$ globules form which have a large solubility for sulfur. Recently, in the U. S., desulfurization by injection of prefluxed $2CaO \cdot Al_2O_3$ has been introduced with some commercial success [reaction (3.7)]. Although these reagents have lower unit cost that carbide or magnesium, there is a limitation shared with lime systems; the greater mass of reagent needed increases the time required for treatment and for the follow on raking step.

Two other methods for delivery of desulfurization agents into hot metal transfer ladles are worthy of note. One approach, paralleling the use of cored wires for steel ladle treatment, is to feed magnesium cored wire at high rates to reach the ladle bottom for release of the reagent at maximum depth. Magnesium in this form is far costlier than as an injectable solid, albeit the delivery system is simpler to operate and maintain. Another approach, with somewhat larger implementation known by the commercial name ISID, consists of feeding the reagents through a rotatable bayonet system installed low in the wall of the hot metal transfer ladle (Fig. 3.12). Maintenance concerns and cost have limited broad implementation of this technology.

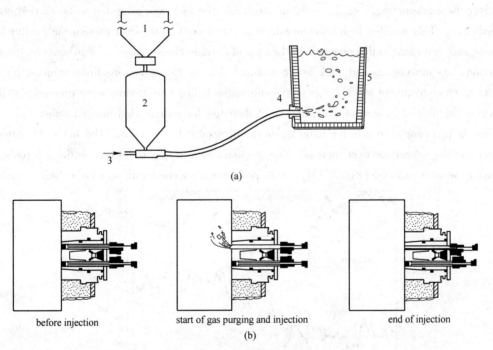

Fig. 3.12 The ISID powder injection process
(a) Equipment schematic; (b) Treatment sequence
1—storage silo; 2—powder feed vessel; 3—transport gas; 4—side injection device; 5—ladle

3.3.3 Transport Systems

Delivery and use of pulverized desulfurization reagents, e.g. magnesium, lime, calcium carbide, entail distinct technological requirements, principally the avoidance of contact with air. Magnesium powder, produced by atomization or grinding, must be transported in sealed, air-tight containers of limited capacity (20,000 kg each). Thus, to provide the capability to move this material in bulk, the industry developed a 90% Mg-10% lime product that is flowable and can be delivered and stored in bulk trailers. Calcium carbide also must be kept from the moisture in air and is transported and stored in bulk, sealed, pneumatic system equipped trailers. Salt-coated magnesium is relatively impervious to moisture and generally is stored in bulk trailers as well.

An important adjunct to facilitate the use of pulverized materials such as lime and calcium carbide has been the development of a technology for improvement of the flowability of these materials by application of silicone oil based flow aids during pulverization. Powders prepared in this manner can be delivered by dense phase injection techniques, which minimize the amounts of reagent and iron droplets carried out of the liquid by the transport gas (e. g., for lime, transport line loading of 2kg/L of gas). This allows delivery of injection reagents at rates of 50kg/min through an 18mm transport line.

3.3.4 Process Venue

When the pneumatic delivery of reagents was first introduced in the 1970s, these processes were relatively time consuming, i. e. 20 to 30 minutes, and the task was relegated to be carried out in torpedo cars. This resulted in interference from slag reactions as the ever present high sulfur blast furnace slag can cause sulfur reversals in the case of carbide based reagents. Furthermore, the postreaction slags are viscous and stick to the roof and sides in the submarine fleet, reducing holding capacity. Over-treatment was the rule as the submarine ladles were treated some time prior to their arrival at the melt shop and the coordination of their use for an intended product order.

Most shops changed to transfer ladle treatment to resolve these issues. The hot metal carrying capacity of the submarine fleet increases due to elimination of slag build ups on the sub roofs and refractory wear is reduced (Fig. 3.13). Other problems associated with submarine ladle treatment

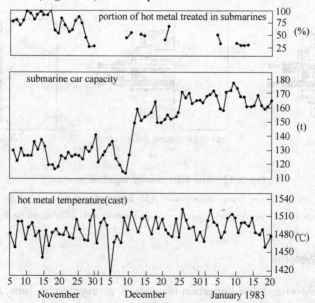

Fig. 3.13 Illustration of gain in submarine (torpedo) car capacity upon cessation of desulfurization treatment (lime-Mg)

include relatively poor mixing caused by the shape of the elongated bottle configuration and the relatively shallow liquid depth. In the BOF transfer ladle, reagent efficiency of magnesium is increased by the greater depth of lance immersion, which provides longer residence time for the magnesium bubbles to travel to the surface. The major benefit of treatment in the transfer ladle rather than in the submarine ladle is that it provides the opportunity to treat individual hot metal charges to specific sulfur levels set by the requirements of the intended steel grade.

3.3.5 Slag Management

As in all metallurgical processes, management of the slag produced during hot metal desulfurization is critical to success. After conclusion of treatment, the slag usually is removed with a raking device, which typically is an articulated arm and paddle assembly. The raking process requires some time which may become a production penalty in some operations. Process yield suffers as some hot metal is lost from the ladle with each stroke of the paddle. In some shops, the iron transfer ladle has a retention dam across the mouth with hole(s) for the metal to pour out. The slag is retained in the ladle as the hot metal is charged into the furnace. While effective at separating the slag and minimizing yield loss, this may slow the rate of charging the vessel and, therefore, extend the converter heat cycle. Additionally, the slag retained in the ladle must be dumped after each use; as this is done by reversing direction (to keep the pour holes open), this step may create difficulties in some shops.

During raking the post-treatment slag will take with it a significant metallic content (approximately 40% by weight), this can represent a yield loss of nearly 1%. Methods to minimize this loss include the use of dense phase injection (to minimize the volume of gas for delivery of the injected powders), and the addition of a fluxing agent, about 5% CaF_2 or Na_2CO_3 to the desulfurizing reagents, to produce a less viscous liquid slag for release of the iron globules. A further help is to provide a small amount of gas bubbling (Fig. 3.14) during the raking process; this promotes flotation of the slag towards the lip of the ladle and thereby reduces the number of strokes required for slag removal.

Typical figures for slag removal are in the range of 15-25kg/t hot metal for most U.S. shops. Two viscosity related factors combine to make the quantity of slag raked from transfer ladles dependent on hot metal temperature, Fig. 3.15: colder hot metal appears to "hold" more entrained blast furnace slag and the retention of iron droplets in slags increases as temperature drops. The trend shown in Fig. 3.15 is typical for most BOF shops. Disposal of the spent slag usually is by mixing it into the blast furnace slag management system despite the remaining unused sulfur holding capacity. In some plants special controls are in place to cope with the effect of remaining unreacted reagents such as carbide or soda ash. This is not an issue with magnesium or lime.

3.3.6 Lance Systems

Fig. 3.16 illustrates the commonly used injection lance designs. Nozzles may be directed vertically or exit the side of the lance at various angles. Typical life experience is 80 treatments or up to 1200 minutes for the hockey stick design, which prevents the reactive gases from attacking the re-

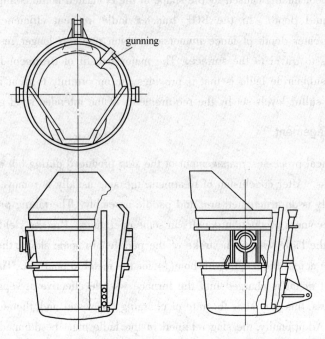

Fig. 3.14 Illustration of a transfer ladle slag stirring lance

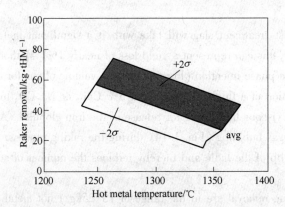

Fig. 3.15 Effect of hot metal temperature on raking loss after desulfurization with lime-Mg

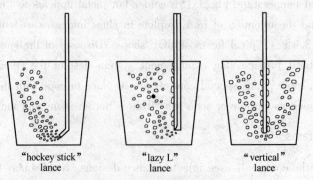

Fig. 3.16 Commonly used injection lance configurations

fractory coating. A lower figure, 70 treatments, is typical for the simpler lazy L lances.

Lances typically are constructed of lengths of square steel tubing which contain the transport pipe(s). These assemblies are then cast within a refractory mold and cured in temperature controlled drying ovens. The cross sectional shape may be square or round with an area of 40-50cm^2 (6-8 in^2). The refractory typically is high alumina.

A recent development directed at increasing the injection rate of magnesium containing combinations, without causing otherwise intolerable violence at the surface of the metal in the transfer ladle, is to use two transport lines contained within a single lance each fed by a separate injection materials source. The outlet nozzles are positioned to deliver the reagent at 30° from the vertical on opposite sides of the lance. It is also possible to use two separate lances immersed into the metal at the same time. These approaches have reduced injection time by 40%-50% without increasing iron losses to the ladle slag or reduction in reagent efficiencies.

3.3.7 Cycle Time

In a typical BOF shop, the hot metal treatment operation can become a limit on productivity. Pour out and movement of the submarine ladles may require 5-10 minutes. Injection time will consume 7-20 minutes, depending upon the amount of sulfur to be removed, followed by raking time of 5-10 minutes.

3.3.8 Hot Metal Sampling and Analysis

Hot metal sampling and analysis is a critical issue, particularly for magnesium reliant systems, because time must be provided for removal by flotation of the MgS reaction product. Best reproducibility in results is obtained when the post-treatment sample is obtained after the raking step is completed (Fig. 3.17). Similarly, gas stirring systems installed to aid the raking operation (Fig. 3.14), can result in improved sampling accuracy as both the pre-and post-treatment samples are made more representative.

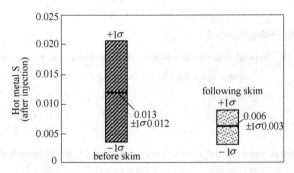

Fig. 3.17 Decrease in variability of after-treatment analyses (lime-Mg process)

The method for determination of endpoint sulfur is, in part, a function of the precision required at the individual melt shop and other operating considerations. Gas combustion methods (Leco) and optical emission spectrometers (OES) are used commonly. Many shops have these instruments at the treatment sites for rapid determination of results.

Optical emission spectrometers have demonstrated sufficient precision for determination of ppm concentrations of sulfur in hot metal provided the disk samples obtained for analysis are sound and prepared properly. Pits, holes and cracks will affect the results, as will surfaces with improper texture and/or flatness. OES determination has the advantage of providing silicon and manganese analysis for use in the furnace charge model—a significant benefit when erratic blast furnace operations result in varied hot metal chemistry.

The gas combustion method is recognized as probably the most accurate for determination of sulfur (and carbon) due to ease in achieving appropriate sample integrity. Prevention of process slag entrapment in the pin sample is the greatest concern. Properly designed immersion samplers inserted quickly to sufficient depth will provide slag-free samples. The latest Leco C-S determinator is advertised to have a precision of 0.15ppm.

3.3.9 Reagent Consumption

Reagent consumptions for lime plus magnesium and carbide plus magnesium process systems are illustrated in Fig. 3.18. Control of powder flowability and rates is critical for achievement of these consumption levels. It is essential to avoid excess transport gas and/or visible magnesium flares or powder plumes.

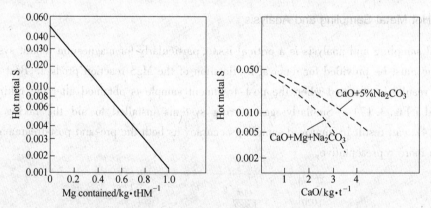

Fig. 3.18 Typical reagent consumptions for desulfurization by co-injection.
Left: lime-Mg (lime containing 6% spar)

3.3.10 Economics

Fig. 3.19 presents a breakdown of the cost components for a typical transfer ladle hot metal desulfurization operation.

3.3.11 Process Control

Modern facilities are generally fully automated to control the operation from the start of treatment to finish. Powder initiation and lance insertion are automatically controlled along with lance withdrawal after the predetermined amount of reagent has been injected. Reagent rates may be adjusted

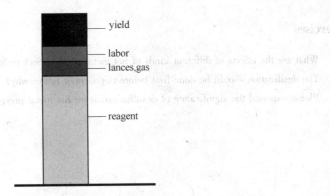

Fig. 3.19 Relative cost components for desulfurization of hot metal.
Yield refers to loss of hot metal during slag raking

during treatment by controlling injection tank pressure or transport line throat diameter. In addition to the automatic functioning, the controlling computer provides for process data capture and storage. Real-time graphics provide information on process efficiency and dispenser performance to aim standards. Typical, for a well-controlled operation, is to achieve within 0.002%S of aim at end points ordered less than 0.003%S.

3.4 Hot Metal Thermal Adjustment

A one-of-a-kind facility to heat hot metal has been installed by ISCOR in South Africa. The 1500 ton hot metal mixer (Fig. 3.20), is equipped with channel inductors which provide induction heating and stirring of the liquid. Options for use are to melt scrap in the unit by energy addition to the hot metal or to raise the hot metal temperature and thereby improve the scrap melting capability of the converter. Superheat of 225°C (440°F) for a 40% scrap rise increase is achievable. Electrical energy conversion for temperature gain has been reported as 85%. Whereas technically successful in performing its functions, the capital cost of such a system, in contrast to the scrap melting benefit obtainable with chemical energy sources, has made this approach unattractive for other steel plants.

Fig. 3.20   1500 tonne superheater channel furnace, rated at 15,000kW

Exercises

3-1 What are the effects of different kinds of hot metal pretreatment technologies?

3-2 The desilication should be done first before dephosphorization, why?

3-3 Please expound the significance of desulfurization for hot metal pretreatment.

4 Oxygen Steelmaking Processes

4.1 Introduction

4.1.1 Process Description and Events

The oxygen steelmaking process rapidly refines a charge of molten pig iron and ambient scrap into steel of a desired carbon and temperature using high purity oxygen. Steel is made in discrete batches called heats. The furnace or converter is a barrel shaped, open topped, refractory lined vessel that can rotate on a horizontal trunnion axis. The basic operational steps of the process (BOF) are shown schematically in Fig. 4.1.

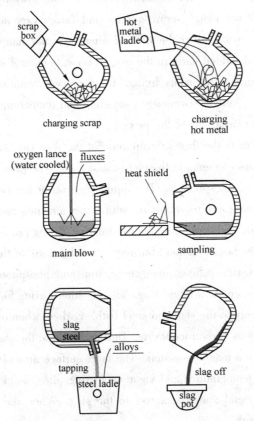

Fig. 4.1 Schematic of operational steps in oxygen steelmaking process (BOF)

The overall purpose of this process is to reduce the carbon from about 4% to less than 1% (usually less than 0.1%), to reduce or control the sulfur and phosphorus, and finally, to raise the temperature of the liquid steel made from scrap and liquid hot metal to approximately 1635℃ (2975°F). A typical configuration is to produce a 250 ton (220 metric ton) heat about every 45 minutes, the range is approximately 30 to 65 minutes. The major event times for the process are summarized below in Table 4.1.

Table 4.1 Basic oxygen steelmaking event times

Event	Time/min	Comments
Charging scrap and hot metal	5-10	Scrap at ambient temperature, hot metal at 1340℃ (2450°F)
Refining-blowing oxygen	14-23	Oxygen reacts with elements, Si, C, Fe, Mn, P in scrap and hot metal and flux additions to form a slag
Sampling-chemical testing	4-15	Steel at 1650℃ (3000°F), chemistry and temperature
Tapping	4-8	Steel is poured from furnace into a ladle, typical size = 250 tons
Pouring slag off at furnace	3-9	Most slag is removed from furnace, in some shops slag is used to coat furnace walls

These event times, temperatures and chemistries vary considerably by both chance and intent. The required quantities of hot metal, scrap, oxygen and fluxes vary according to their chemical compositions and temperatures, and to the desired chemistry and temperature of the steel to be tapped. Fluxes are minerals added early in the oxygen blow, to control sulfur and phosphorous and to control erosion of the furnace refractory lining. Input process variations such as analytical (hot metal, scrap, flux and alloy) and measurement (weighing and temperature) errors contribute to the chemical, thermal and time variations of the process.

The energy required to raise the fluxes, scrap and hot metal to steelmaking temperatures is provided by oxidation of various elements in the charge materials. The principal elements are iron, silicon, carbon, manganese and phosphorous. The liquid pig iron or hot metal provides almost all of the silicon, carbon, manganese and phosphorous, with lesser amounts coming from the scrap. Both the high temperatures of the liquid pig iron and the intense stirring provided when the oxygen jet is introduced, contribute to the fast oxidation (burning or combustion) of these elements and a resultant rapid, large energy release. Silicon, manganese, iron and phosphorous form oxides which in combination with the fluxes, create a liquid slag. The vigorous stirring fosters a speedy reaction and enables the transfer of energy to the slag and steel bath. Carbon, when oxidized, leaves the process in gaseous form, principally as carbon monoxide. During the blow, the slag, reaction gases and steel (as tiny droplets) make up a foamy emulsion. The large surface area of the steel droplets, in contact with the slag, at high temperatures and vigorous stirring, allow quick reactions and rapid mass transfer of elements from metal and gas phases to the slag. When the blow is finished the slag floats on top of the steel bath.

Controlling sulfur is an important goal of the steelmaking process. This is accomplished by first removing most of it from the liquid hot metal before charging and later, inside the furnace, by controlling the chemical composition of the slag with flux additions.

4.1.2 Types of Oxygen Steelmaking Processes

There are basically three variations of introducing oxygen gas into the liquid bath. These are shown schematically in Fig. 4.2. Each of these configurations has certain pros and cons. The most common configuration is the top-blown converter (BOF), where all of the oxygen is introduced via a water-cooled lance. The blowing end of this lance features three to five special nozzles that deliver the gas jets at supersonic velocities. In top blowing, the stirring created by these focused, supersonic jets cause the necessary slag emulsion to form and keeps vigorous bath flows to sustain the rapid reactions. The lance is suspended above the furnace and lowered into it. Oxygen is turned on as the lance moves into the furnace. Slag forming fluxes are added from above the furnace via a chute in the waste gas hood.

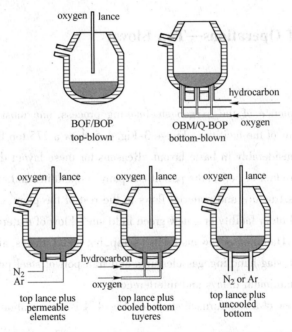

Fig. 4.2 Methods of introducing oxygen and other gases into the steelmaking converter

In the bottom-blown converters (OBM or Q-BOP), oxygen is introduced via several tuyeres installed in the bottom of the vessel (Fig. 4.2). Each tuyere consists of two concentric pipes with the oxygen passing through the center pipe and a coolant hydrocarbon passing through the annulus between the pipes. The coolant is usually methane (natural gas) or propane although some shops have used fuel oil. The coolant chemically decomposes when introduced at high temperatures and absorbs heat in the vicinity, thus protecting the tuyere from overheating.

In bottom blowing, all of the oxygen is introduced through the bottom, and passes through the bath and slag thus creating vigorous bath stirring and formation of a slag emulsion. Powdered fluxes are introduced into the bath through the tuyeres located in the bottom of the furnace.

The combination blowing or top and bottom blowing, or mixed blowing process (Fig. 4.2 shows these variants) is characterized by both a top blowing lance and a method of achieving stirring from the bottom. The configurational differences in mixed blowing lie principally in the bottom tuyeres or elements. These range from fully cooled tuyeres, to uncooled tuyeres, to permeable elements.

4.1.3 Environmental Issues

The oxygen steelmaking process is characterized by several pollution sources and most require emission control equipment. These sources are hot metal transfer, hot metal desulfurization and skimming of slag, charging of hot metal, melting and refining (blowing), BOF tapping, handling of dumped BOF slag, handling of fluxes and alloys and maintenance (burning of skulls, ladle dumping. etc). Thus, compliance to emission standards is an important design and operating cost factor for the operation.

4.2 Sequence of Operations—Top Blown

4.2.1 Plant Layout

To understand the sequence of the oxygen steelmaking process, one must examine the design, layout and materials flow of the facilities. Fig. 4.3-Fig. 4.5 show a 275 ton BOF that illustrates the process. Shops vary considerably in basic layout. Reasons for these layout differences are: type of product (ingots, cast product or both), the parent company's operating and engineering culture, the relationship of the infrastructure and material flows to the rest of the plant, and age of the facility. Is the plant an updated older facility or a new green field site? Flow of materials plays a key role in the design of the shop. Handling of raw materials (scrap, hot metal, fluxes, alloys, refractories), oxygen lances in-and-out, slag handling, gas cleaning, and transport of steel product must be accomplished smoothly with minimum delays and interference.

Fig. 4.3 is a plan view of a two-furnace shop and Fig. 4.4 is an elevation of the same plant but

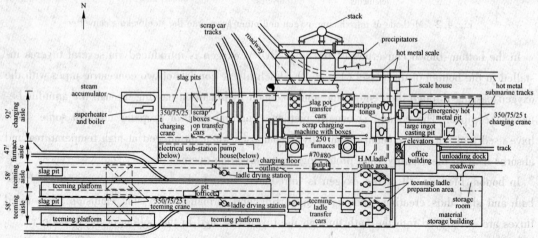

Fig. 4.3 Plan view of 275 ton BOF shop

looking to the west. Fig. 4.5 is an elevation looking to the north. BOFs, OBMs (Q-BOPs) and other variants can have similar layouts except for oxygen conveying and flux handling details. All shops feature transportation systems for hot metal and scrap.

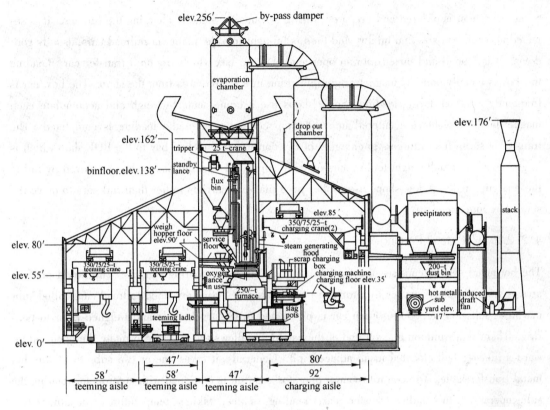

Fig. 4.4 Elevation of 275 ton BOF shop—looking west

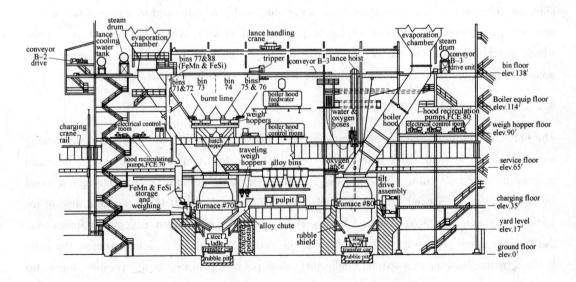

Fig. 4.5 Elevation of 275 ton BOF shop—looking north

4.2.2 Sequence of Operations

4.2.2.1 Scrap Handling

Scrap for a heat is ordered and prepared well in advance of actually charging the furnace. It is selected according to size and quality and then is brought into the plant via railroad cars, usually gondolas. It is loaded and mixed into an open ended scrap box which sits on a transfer car. Loading the box is usually done by magnet or grapple crane in a remote area from the shop. The box/car is frequently weighed during loading. Some shops use a crane scale to weigh and accumulate each magnet load. Weights are entered into the shop computer when the loading is completed. The transfer of scrap from rail cars to charging box is done in an attached bay to the BOF shop which is large enough to handle eight to 24 hours of scrap supply. The scrap box is then conveyed by rail to the charging aisle. A few shops use rubber tired platform carriers rather than rail cars to move the scrap box into the shop.

4.2.2.2 Hot Metal Pouring

The hot metal system consists of a track(s) and one to three pouring stations. The liquid pig iron arrives from the blast furnace in a train of torpedo shaped, refractory lined railroad cars called submarines (subs) or torpedoes. Each car is positioned over a track scale and weighed prior to pouring. There is a trunnion at each end of the car which allows the operator to rotate the open top toward a transfer ladle located in an adjacent pit. Generally, it takes one or two subs to fill the hot metal transfer ladle. The control room or operator's pulpit is equipped with controls for rotating the sub, operating the ladle transfer car, reading scales, taking temperatures, desulfurization equipment, and sending samples to a chemistry lab. The pouring operation, which generates considerable dust emissions, is accomplished under an enclosed hood equipped with an evacuation system and a baghouse. The dust generated at pouring, called kish, is mainly fine flaked graphite which precipitates from the carbon saturated metal as its temperature drops during pouring. The poured weights and measured temperatures are entered into the shop process control computer.

4.2.2.3 Hot Metal Treatment

The hot metal transfer ladle sits on a transfer car at the outside wall of the charging aisle, usually out of the crane's reach. Here many shops treat the hot metal by injecting a mixture of lime and magnesium to remove sulfur. This process is called hot metal desulfurization. During hot metal treatment, sulfur is removed from approximately 0.025% to as low as 0.002% and the time of injection will range from 5-20 minutes. A gas collecting and filtration system collects the fumes from the desulfurization process as well as collecting the pouring fumes. After desulfurization, the ladle is tilted by the crane or in a special cradle just to the point of pouring. In this position, the sulfur containing slag floating on the iron is scraped off into a collection pot using a hydraulic manipulating arm. This slag removal process is called skimming. Often, much metal is scraped out of the ladle along with the sulfur containing slag. Thus there is an iron yield loss due to deslagging that ranges from 0.5% to 1.5% depending on equipment design and operator skill. Hot metal temperature is measured using disposable thermocouples mounted on a mechanical arm called a pantograph. When the hot metal pouring and treatment are finished, the ladle car moves into the charging aisle and the ladle becomes available for pickup and charging by the charging crane.

4.2.2.4 Charging the Furnace

The BOF furnaces are open-topped, refractory-lined vessels located in the adjacent aisle called the furnace aisle. The furnaces rotate on trunnions so they can be tilted both toward the charging and tapping aisles. The furnace refines the steel in an upright position although it is capable of rotating 360°. The furnace aisle contains the furnaces, flux conveying system, furnace alloy chute and oxygen lances in a top-blown shop. Here, the furnace aisle is very high to accommodate loading and operation of 60-70 foot (18-21m) long oxygen lances. There is a space and capital cost advantage of an OBM (Q-BOP) that does not require the high elevation for lances.

In nearly all North American shops, scrap is charged first. Many shops lift and tilt the box emptying the scrap into the furnace with the charging crane. Charging scrap before hot metal is considered a safer practice that avoids splashing. The crane method usually has faster scrap charging times. However, many shops load the scrap boxes onto special charging machines that can move on rails in front of the furnace. This scrap charging machine has a hydraulic tilting mechanism that raises the scrap box to 45° and charges them one at a time. It usually holds two scrap boxes. While the scrap charging machine is often slower than the crane, it frees up the charging crane for other duties and more quickly handles two box charges per heat when required.

After scrap is charged, liquid hot metal is charged into the furnace using the charging crane in the charging aisle. The ladle is tilted and the liquid hot metal is poured into the furnace. This process takes 1-5 minutes depending on the design of the furnace hood and shop fugitive emission systems. Some shops can charge quickly because the fume from pouring into the furnace is effectively collected by the hood and a closed roof monitor collection system. Other shops with less advanced fume collection systems, must pour rather slowly to minimize the heat and fume cloud, thus taking a lengthy 3-5 minutes to charge hot metal.

4.2.2.5 Computer Calculations

Prior to pouring hot metal and charging scrap, a computer calculation is initiated by the pulpit operator to determine the charge recipe. The grade and temperature and chemistry aims are loaded into the shop computer beforehand when a heat is put on the schedule line-up. The temperature and chemical content of the hot metal can vary significantly. The hot metal is sampled and analyzed at the chemistry lab—a process that takes from 3-10 minutes—and the results and ID are transmitted to the computer. The temperature of the hot metal is measured in the ladle after it is poured and that result is transmitted to the computer. The chemistry of the scrap is calculated from the known mixture of the scrap; its temperature is assumed to be ambient. Thus, all of the characteristics of the charged materials and the heat aims are available for the charge recipe calculation. The principal aim parameters for the furnace heat are carbon and temperature. Other specific aims are sulfur, phosphorous and slag composition (%FeO level). Missing any of these aims can be very costly later and can require time consuming corrective actions or create significant quality problems. Often, there are several update calculations as items are weighed and charged to correct for minor weighing irregularities and mistakes. The results of these calculations are principally the amounts of coolants, fluxes and oxygen. Most shops' calculations have features that allow early calculation of hot metal and scrap weights as well. All shops use the charge recipe calculation. Some shops have sensors (sub-lance, light meter, bomb-drop thermocouples, etc.) that measure carbon and/or temperature near the end of the blow. The sensor's measurements are used to make late oxygen and coolant corrections to bring the carbon and temperature to aims. This saves time and money by reducing the frequency of corrective actions.

4.2.2.6 Oxygen Blow

After scrap and hot metal are charged, the furnace is set upright and the oxygen is supplied through a water-cooled lance. There are two lance lift carriages above each furnace but only one lance is used at a time; the other is a spare. The oxygen blow times typically range from 13 to 25 minutes from one shop to another with an average of about 20 minutes. The oxygen is added in several batches. Each batch is characterized by a different lance height above the static steel bath and sometimes by an oxygen rate change. These blowing rates and lance heights vary considerably from shop to shop and depend on the pressure and quality of the oxygen supply. The oxygen blow rate ranges from 560 to 1000 Nm^3 per minute (20,000 to 35,000 scfm). A practical limit on the rate is often the volume of the furnace and the capacity of the gas collection and cleaning system to handle the gaseous reaction product and fume. A typical example of the oxygen batches is summarized in Table 4.2.

Table 4.2 Example of oxygen batches in a BOF

Batch No.	Lance Ht/inch(cm)	Oxygen volume at lance change/Nm^3
1	150(381)	850
2	120(305)	1700
3 (main)	90(229)	balance (to approx 14,200)

The first batch lance height is very high to avoid the possibility of lance tip contact with the scrap and to safely establish the oxidizing, heat generating reactions. If the lance would contact the pile of scrap in the furnace, a serious water leak could result causing a dangerous steam explosion.

The second batch lance height is usually approximately 20-30 inch (50.8-76.2cm) lower than the first batch and approximately 20-30 inch (50.8-76.2cm) higher than the main batch. The purpose here is to increase the reaction rate and control the early slag formation. This second or middle batch generates some early iron oxide to increase proper slag formation.

The main batch is where most of the action occurs—it is by far the longest batch. The lance height is an empirical compromise between achieving faster carbon removal rates and proper slag making. Some shops have more than three batches. Some change oxygen conditions (blow rate and lance height) nearly continuously. Other shops will raise the lance and change the blow rate near the end of the main batch to control the viscosity and chemical reactivity of the slag by raising its FeO content.

The position of the lance is very important for proper functioning of the process. If the lance is too high, the slag will be over stirred and over-oxidized with higher FeO percentages. This will cause higher than normal yield losses and lower tap alloy efficiencies due to oxidation losses. Further, the rate of carbon removal is reduced and becomes erratic. Slag volume increases and there is an increased chance of slopping, which is an uncontrolled slag drooling or spilling over the top of the furnace. When the lance is too low, carbon removal increases somewhat, slag formation, slag reactivity, and FeO are reduced and sulfur and phosphorus removal problems often occur. If the lance is very low, then spitting of metal droplets or sparking occurs which cause severe and dangerous metallic deposits, called skulls, on the lance and the lower waste gas hood.

Obviously, there is a correct lance height. It varies from shop to shop and depends on furnace configuration, lance configuration and oxygen supply pressure or flow rate. Each shop must find its own best lance height and comply with it. The problem is how to measure it. It can change quickly and significantly as a result of changes in furnace refractory shapes. Traditional measurement methods have been burn-off tests, where a pipe is wedged in an oxygen port. The lance is then lowered and the pipe is allowed to melt off at the slag-metal interface. This is done while waiting for a chemistry just before tap. This test has been erratic due to the temperature and fluidity of the slag, time of immersion, and initial protection of the pipe. This test has fallen out of favor because it is dangerous for the operator to attach the pipe to the lance.

More modern techniques include mathematically integrating the furnace volume from a refractory lining laser scan or determining the distance to the bath/slag using a radar unit mounted above the furnace. Generally, the radar method measures the height of the slag surface after it has collapsed at the end of a low carbon blow. There is uncertainty about the location of the slag steel interface but the measurement is considered better than none.

4.2.2.7 Flux Additions

Soon after the oxygen is turned on, flux additions are started and are usually completed at the end of the second batch of oxygen. The fluxes control the chemistry and sulfur and phosphorus capacity of the slag. The principle active ingredients from the fluxes are CaO (from burnt lime) and MgO (from dolomitic lime). The CaO component is used principally to control sulfur and phosphorous.

The dolomitic lime is used to saturate the slag with MgO. The principle ingredient of the furnace refractories is MgO. Steelmaking slags without it are very corrosive to the lining. The corrosion rate is reduced dramatically when MgO is added to saturate the slag. It is much cheaper to satisfy the slag's appetite for MgO from dolomitic lime than by dissolving it from the lining.

Another flux addition sometimes used in high carbon heats is fluorspar (CaF_2 or spar). This mineral is charged to dissolve the lime and to reduce the viscosity of the slag. It is used for making high carbon heats ($w[C]$ 0.30% at the end of blow), because the iron oxide concentrations are low on these heats. Iron oxides help to dissolve lime in lower carbon heats, but these oxides are present in low concentrations in high carbon heats. To compensate for less FeO, many shops use spar to dissolve the lime. However, spar is used very sparingly because it is very corrosive on refractory linings. Unfortunately there is no corrosion inhibiting practice or ingredient to stop the corrosive effects of spar. In addition, spar forms hydrofluoric acids in the gas cleaning system that seriously corrode any metal surfaces in the hood and cleaning systems. Finally, significant fluoride emissions are serious pollution and health hazards.

Coolants are other additions often made at about the same time as fluxes. There are several types of coolants. Iron ore, either lump or pellets, are the most common type. Varieties of limestone (calcium and/or magnesium carbonates) are often used but the cooling effects are less dependable than ores. Some shops use pre-reduced pellets which contain about 93% iron and thus behave similar to scrap. The coolant amounts are calculated by the computer. Ore (iron oxide) should be added as soon as possible to achieve early lime dissolution and to reduce the possibility of vigorous reactions and slopping at mid blow.

4.2.2.8 Final Oxygen Adjustments and Dynamic Sensors

The third or main batch is usually blown at 80-95 inch (203.2-241.3cm) lance height above the bath depending on furnace design, practice and available oxygen pressure.

In many shops, the oxygen lance height is changed near the end of the blow to control the iron oxide (FeO) in the slag. Some shops activate their dynamic control systems to measure carbon and temperatures at this point. Oxygen is turned off based on either the static charge calculation or based on a modified result calculated from the dynamic sensor(s).

4.2.2.9 Turndown and Testing

After the blow is finished, the furnace is then rotated towards the charging side. Often the slag is very foamy and fills up the upper volume of the furnace. This foam will often take several minutes to collapse and settle down on its own. Thus, the operators will often toss pieces of wood or cardboard or scrapped rubber tires onto the bath to increase the collapse of the foam.

There is a heat shield on the charging floor on a track that is positioned in front of the furnace during the blow. The mouth of the furnace is rotated toward the charging side nearly 90°, so the operator can look inside the furnace and sample the heat for chemical analysis and temperature measurement. Here, he also assesses the furnace condition to determine when and if any special maintenance is required.

4.2.2.10 Corrective Actions

Based on the chemical laboratory results, the melter decides if the heat is ready for tap or requires corrective action—a reblow and/or coolant. If a corrective action is required, the furnace is set upright. A reblow of additional oxygen may be required, with or without coolants or fluxes, to arrive at the desired (aim) chemistry and temperature. Usually, after a corrective action, another furnace turndown is required, adding 5-8 minutes to the heat time. When the heat is ready, the furnace is rotated upward and over toward the tap side.

4.2.2.11 Quick Tap Procedures

Japan and some European shops reduce sampling and testing times to 1-3 minutes by using a quick tap procedure. Most of these shops use sub-lances to measure temperature and carbon by the liquidus thermal arrest technique. This testing is done without moving the furnace from the upright position. Success of quick-tap depends on consistently meeting the sulfur and phosphorus specification. This procedure can save 3-6 minutes of lab analysis time. These shops simply proceed immediately to tapping the furnace. Consequently, such shops often turn out 60 heats per day from two active furnaces.

Some North American shops have adopted a simplified variation of the quick-tap practice. A few have sub-lances. Others use the bomb drop-in thermocouples with or without oxygen sensors.

Here a heavy cast iron bomb assembly with a specially wound and protected lead wire is dropped into the furnace. The wire lasts long enough to get a reading. The readings are more accurate if the oxygen is stopped, but some shops get a usable reading during the blow. Again, tramp elements, S, P and other residuals are assumed to be acceptable and tapping proceeds immediately. Some two furnace shops use this technique to minimize production losses when one of the furnaces is being relined or repaired.

4.2.2.12 Tapping

For tapping, the furnace is rotated to the tap side and the steel flows through a taphole into a ladle sitting on a car below. The slag floats on top of the steel bath inside the furnace. Near the end of tapping (4-10 minutes) a vortex may develop near the draining taphole and entrain some of the slag into the ladle. There are various devices used to minimize or detect the onset of slag. Heavy uncontrolled slag entrainment into the ladle has a significant adverse effect on production costs and steel quality.

During tapping, alloys are added to adjust the composition to the final levels or to concentrations suitable for further ladle treatment processes. Typically, 2000-6000lbs (907-2722kg) of alloys are added at tap. After tap, the ladle may be transported for further processing to a ladle arc furnace and/or a degasser. Some shops and some grades permit transport to ingot teeming or to the caster without any further treatments.

An increasing number of grades require limiting the amount of slag carryover to the ladle and close control of slag viscosity and chemical composition. Various devices have been developed to minimize slag draining from the furnace. There are two main techniques. One method consists of slowing down the pouring stream at the end of tap with a refractory plug. Usually a ball-shaped device, called a ball, or a cone shaped device, called a dart, is dropped into the taphole using a carefully positioned boom near the end of tap. These devices have a controlled density, between steel and slag, causing them to float at the slag-steel interface. Thus, it plugs the tap at about the time steel is drained out. These units can be very erratic depending on the geometry of the furnace, device shape and slag characteristics. But many shops have successfully minimized slag carryover into the ladle. Another approach is to detect slag carryover with a sensor coil installed around the taphole refractory. With suitable instrumentation, this system gives the operator an accurate and early warning of slag draining through the taphole at which time tapping is stopped by raising the furnace. The net result of these slag control/detection practices is to reduce furnace slag in the steel ladle, thereby improving chemical consistency and reducing the extent of post-tapping treatments and additions.

The condition and maintenance of the taphole and the furnace wall around it can influence alloy recovery consistency and metallic yield. Poor taphole maintenance and practice can lead to a burn through in either the furnace shell or the taphole support frame. A very small taphole can significantly increase the tap time, reducing productivity, steel temperature and nitrogen pickup in the ladle. A very large taphole will not allow enough time to add and mix the alloy additions in the ladle. Further, aged tapholes have ragged streams with higher surface areas that will entrain air, which in turn dissolves more oxygen and makes control of oxygen levels in the steel difficult.

A newly installed taphole yields a tap time of 5 or 6 minutes. Tapholes are generally replaced when the tap time falls below 3 minutes. A very important aspect of the melter/operator's job is to carefully monitor the condition and performance of the tap hole.

Steel is often lost to the slag pot, a yield loss, when a pocket or depression develops near or around the tap opening. Such a depression can prevent several tons of steel from being drained into the ladle. Again, the operator must carefully monitor yields and furnace condition and make repairs to prevent this problem.

4.2.2.13 Slagging Off and Furnace Maintenance

After tapping, the furnace is rotated back upright to prepare for furnace maintenance. The remaining slag is either immediately dumped into a slag pot toward the charging side or it is splashed on the walls of the furnace to coat the lining and thereby extend its life. This slag splashing (coating) maintenance is done by blowing nitrogen through the oxygen lance for 2-3 minutes. Often, the furnace is simply rocked back-and-forth to coat or build up the bottom, charge and tap pad areas. Dolomitic stone or dolomitic lime additions are made to stiffen the slag for splashing or to freeze the slag to the bottom. Frequently, repairs by spraying a cement-like refractory slurry are made as required prior to the next charge.

The furnace is then ready for the next heat.

4.2.3 Shop Manning

4.2.3.1 Introduction

Another way of looking at the steelmaking process is to look at how the operators and workers run the process. The shop crew is very much a coordinated team. Virtually all oxygen steelmaking shops operate around the clock—21 shifts per week. Some are able to schedule a down or repair turn once per week or every other week in conjunction with a similar scheduled blast furnace outage. Similarly, the down stream caster facility will also schedule a repair outage.

When manning is summarized, the manning levels are expressed on a per turn basis. To get the totals for 21 turns per week (3 turns per day times 7 days per week) multiply the turn manning by four and add the management, staff and maintenance forces (who work only 5 day turns).

The example here will cover a two furnace shop producing about 40 to 45 heats per day. Three furnace shop issues and exceptions will be noted in the discussion. The actual figures in a given shop may vary from the example presented here as different shops have different numbers of furnaces (two or three) and different demands (steel grades and production levels).

4.2.3.2 Hot Metal

Transporting the hot metal subs to the BOF requires one man if the engine has radio control. The pouring station has an operator who weighs and pours and an assistant who plugs in electric power to tilt the subs and moves the train locally and changes the subs at the pouring station. If there is desulfurizing, there will be one or two other operators to desulfurize, skim and maintain the equipment. A three-furnace shop will double this crew to operate two or three pouring and treatment stations.

4.2.3.3 Charging Crane

A charging crane operator is required, often with a relief operator. Busy shops may operate two charging cranes with a relief operator serving the charge and scrap cranes. Often one relief crane operator serves both the charging and teeming or the casting crane.

4.2.3.4 Scrap

Scrap is often premixed by a contractor or a plant crew at a separate facility nearby. Gondolas bearing scrap are transported into the BOF by the plant railroad. There will be one or two scrap crane operators (depending on how many heats the shop makes) loading and weighing the boxes. The charging crane then picks up the scrap boxes and charges them into the furnace or puts them on a charging machine.

4.2.3.5 Furnace and Charging Floor

The overall responsibility for production of the shop during the turn belongs to the turn supervisor. There is a furnace crew consisting of a melter (who is the crew leader and is responsible for making the heat), pulpit operator who operates the computer, lance and flux systems, and two or three furnace operators. In small, lean shops, the melter is also the turn supervisor. The furnace operators actually position the furnace for charging, (operate a scrap charging machine if there is one) and test the heat by sampling the heats for chemistry and temperature. They also weigh up alloys, tap the heat and maintain the furnace refractory—all under the melter's supervision. In larger, high productivity shops, there may be one of these crews and a melter for each active furnace. In this case, the turn supervisor coordinates all shop activities.

4.2.3.6 Flux Handling

There is a track-hopper operator who controls equipment that unloads the fluxes and alloys from rail cars and trucks. Often there is an assistant at the unloading station at ground level, with the operator on the top floor where the conveyer dumper is located. Generally, this operation occurs on day turn in smaller shops or day and afternoon turns in larger shops.

4.2.3.7 Maintenance

There is an on-turn maintenance crew which responds to delay and breakdown calls and performs certain preventative maintenance procedures that are safe during furnace operation. Usually, the day time crew is larger to handle calls and do preventative maintenance. The maintenance crew also load and unload oxygen lances and maintain the oxygen system. Turn maintenance people consist of mechanical millwrights and electricians (3-10 millwrights and electricians). The day turn crew also includes instrument calibration, instrument repair and electronic person skills.

4.2.3.8 General Labor

Often, there is a small day turn labor crew (one to three persons) which operates a forklift trucks to transport various alloy additions for tapping and assists getting spray materials ready for furnace maintenance. The operators also run a mechanical sweeper and hand sweep around the shop to maintain good housekeeping.

4.2.3.9 Chemistry Lab

In older shops there is a vacuum conveyer systems where samples are transported to the chemistry lab. Laboratory manning will vary widely depending on how centralized the lab is to the whole plant. An increasing number of shops are installing robot labs for faster and more convenient service of chemical analysis. However, the robot lab requires an operator whose responsibility is supervising calibration, changing grinding media, and lubrication and maintenance of mechanical and optical equipment. The advantages of a well maintained robot lab are speed and consistency in sample preparation and analysis rather than labor savings.

4.2.3.10 Refractory Maintenance

Often furnace maintenance is done by the floor furnace crew. However, larger and busier shops will have an outside contractor or a separate crew to spray repair the furnaces. This crew may operate only during day turns and consists of two or three persons.

4.2.3.11 Relines and Major Repairs

Maintenance may schedule a crew to make hood and furnace shell repairs while relining is going on. Often, special code welding skills (usually hood or thick shell repairs) are required and are often done by an outside contractor. With the growth of slag splashing, which causes infrequent relines, hood repair work must be scheduled several times between relines.

4.2.3.12 Ladle Liners

Hot metal and steel ladles are a very important component of steelmaking. Hot metal ladle life is typically 1500 pours and a typical small shop will carry a fleet of four or five ladles of which two or three are active. Steel ladles last from 60 to 150 heats. Steel ladle fleets may vary from 8 to 25 depending on the size of the shop. Relining of ladles is usually supervised by the Teeming Supervisor. Ladle lining crew sizes vary widely.

4.2.3.13 Management and Clerical

Finally, management and clerical staffs vary widely from shop to shop. The General Supervisor (sometimes called Area Manager) supervises the day-to-day operation and plans the course of the facility. Often there are separate area supervisors (five day turns per week only) covering hot metal pouring and treatment, furnace floor operations, scrap and flux loading, and maintenance. Usually there is a small clerical staff (1-5), a practice engineer, and a small, assigned crew of metallurgical or quality assurance (QA) people.

In summary, even in a small shop, the steelmaking team brings together many different skills working in concert. In addition to the production crew, there are numerous support groups, principally maintenance, that keep the shop running smoothly. Table 4.3 summarizes the manning of a small BOF shop.

Table 4.3 Example of manning a small two furnace BOF shop

Position	Area	Turns	Number
General supervisor	Supv	Day	1
Hot metal supervisor	Supv	Day	1
Furnace supervisor	Supv	Day	1
Scrap and flux supervisor	Supv	Day	1
Maintenance supervisor	Supv	Day	1
Melter	Supv	All	4
Maintenance supervisor	Turn	All	4
Hot metal pourer	Turn	All	4
Hot metal assistant	Turn	All	4
Desulf and skim	Turn	All	4
Charging crane	Turn	All	8
Charging crane relief	Turn	All	4

Continued Table 4.3

Position	Area	Turns	Number
Pulpit operator	Turn	All	4
Furnace operator	Turn	All	4
Scrap crane	Turn	All	4
Track hopper	Turn	Day	2
Track hopper assistant	Turn	Day	2
Labor	Turn	All	4
Maintenance	Turn	All	Varies
Ladle liners		Day	Varies
Metallurgist	Staff	Day	2
Clerical	Staff	Day	2

4.3 Raw Materials

4.3.1 Introduction

The basic raw materials required to make steel in the oxygen steelmaking process include: hot metal from the blast furnace, steel scrap and/or any other metallic iron source (such as DRI), ore (Fe_2O_3), and fluxes such as burnt lime (CaO), dolomitic lime (CaO-MgO), dolomitic stone ($MgCO_3$-$CaCO_3$) and fluorspar (CaF_2).

Scrap, charged from a scrap box, is the first material to be charged into the furnace. The hot metal is then poured into the vessel from a ladle, after which the oxygen blow is started. The fluxes, usually in lump form, are charged into the furnace through a bin system after the start of the oxygen blow. The fluxes can also be injected into the furnace in powder form through bottom tuyeres.

The composition and amounts of raw materials used in the steelmaking process vary from one shop to another, depending on their availability and the economics of the process. The basic raw materials used in the oxygen steelmaking process are described below.

4.3.2 Hot Metal

The hot metal, or liquid pig iron, is the primary source of iron units and energy in the oxygen steelmaking process. Hot metal is usually produced in blast furnaces, where it is cast into submarine shaped torpedo cars and transported either to a desulfurization station or directly to the steelmaking shop.

4.3.2.1 Composition

The chemical composition of hot metal can vary substantially, but typically it contains about 4.0%-4.5% carbon, 0.3%-1.5% silicon, 0.25%-2.2% manganese, 0.04%-0.20% phosphorus and 0.03%-0.08% sulfur (before hot metal desulfurization). The sulfur level in desulfurized hot metal can be as low as 0.001%. The composition of the hot metal depends on the practice and charge in the blast furnace. Generally, there is a decrease in the silicon content and an increase in the sulfur of the hot metal with colder blast furnace practices. The phosphorus contents of the hot metal increases if the BOF slag is recycled at the sinter plant.

Carbon and silicon are the chief contributors of energy. The hot metal silicon affects the amount of scrap charged in the heat. For example, if the hot metal silicon is high, there will be greater amounts of heat generated due to its oxidation, hence more scrap can be charged in the heat. Hot metal silicon also affects the slag volume, and therefore the lime consumption and resultant iron yield.

4.3.2.2 Determination of Carbon and Temperature

The hot metal is saturated with carbon, and its carbon concentration depends on the temperature and the concentration of other solute elements such as silicon and manganese. The carbon content of the hot metal increases with increasing temperature and manganese content, and decreases with increasing silicon content.

It is important to know the temperature and the carbon content of hot metal at the time it is poured into the BOF for steelmaking process control. The hot metal temperature is normally measured at the hot metal desulfurizer or at the time it is poured into the transfer ladle from the torpedo cars. If the hot metal temperature has not been measured close to the time of its charge into the BOF, then it can be estimated using the last hot metal temperature measurement, in conjunction with a knowledge of the rate of the hot metal ladle temperature loss with time, and the time elapsed between the last temperature measurement and the BOF charge. Typically, the temperature of the hot metal is in the range of 1315-1370°C (2400-2500°F). Once the temperature is measured or calculated, the carbon content of the hot metal at the time of charge can be estimated using a regression equation, which is mainly in terms of the temperature, hot metal silicon and manganese. Calculating the carbon in this manner has turned out to be as accurate as analyzing it chemically while saving considerable lab effort.

4.3.2.3 Hot Metal Treatment

Desulfurization is favored at high temperatures and low oxygen potentials. Also, the presence of other solute elements in the metal such as carbon and silicon increases the activity of sulfur, which in turn enhances desulfurization. Thus low oxygen potential and high carbon and silicon contents make conditions more favorable to remove sulfur from hot metal rather than from steel in the BOF.

Not all hot metal is desulfurized. Hot metal used for making steel grades with stringent sulfur specifications is desulfurized in the hot metal desulfurizer. The hot metal is poured into a transfer ladle from a torpedo car. It is then transported to the desulfurization station where the desulfurizer can reduce hot metal sulfur to as low as 0.001%, but more typically to 0.004% or 0.005%.

Typical desulfurizing reagents include lime-magnesium, and calcium carbide. Powdered reagents are generally injected using nitrogen gas. Apart from reducing sulfur to low levels, a hot metal desulfurizer can also allow the blast furnace operator to increase productivity by reducing the limestone burden and thereby producing higher sulfur hot metal.

It is important that the slag produced after hot metal desulfurization is removed effectively through slag skimming. This slag contains high amounts of sulfur, and any slag carried over into the BOF, where conditions are not good for desulfurization, will cause sulfur pickup in the steel.

4.3.2.4 Weighing

The weighing of the hot metal is done on a scale while it is being poured into the transfer ladle. It is very important that the weight of the hot metal is accurately known, as any error can cause problems in turndown chemistry, temperature and heat size in the BOF. This weight is an important input to the static charge model.

4.3.3 Scrap

Scrap is the second largest source of iron units in the steelmaking operation after hot metal. Scrap is basically recycled iron or steel, that is either generated within the mill (e.g. slab crops, pit scrap, cold iron or home scrap), or purchased from an outside source.

The scrap is weighed when loaded in the scrap box. The crane operator loads the box based on the weight and mix requirements of the upcoming heat. Then the box is transported to the BOF. It is important that the crane operator loads correct amounts and types of scrap (the scrap mix) as indicated by the computer or a fixed schedule. Otherwise the turndown performance of the heat will be adversely affected. Some typical types of scrap used in a BOF heat, and few of their properties are listed in Table 4.4.

Table 4.4 Types of scrap used in the BOF and their characteristics and chemistry

Type of scrap	Melting	Relative cost	Bulk density /cwt·m^{-3} (45.36kg·m^{-3})	Yield /%	$w(Fe)$ /%	$w(C)$ /%	$w(Si)$ /%	$w(Mn)$ /%	$w(P)$ /%	$w(S)$ /%	$w(Cr)$ /%	$w(Cu)$ /%	$w(Ni)$ /%	$w(Mo)$ /%	$w(Sn)$ /%
Plate and structure	Easy	Moderate	45	94.6	95.5	0.25	0.10	0.10	0.025	0.025	0.09	0.13	0.09	0.02	0.025
Punching and plate	Easy	Moderate	50	94.0	96.0	0.20	0.10	0.30	0.015	0.025	0.06	0.09	0.06	0.01	0.008
No.1 heavy melting	Easy	Average	50	93.3	94.5	0.25	0.10	0.30	0.020	0.040	0.10	0.25	0.09	0.03	0.025
No.2 bundles	Easy	Inexpensive	50	88.0	90.0	0.25	0.10	0.30	0.030	0.090	0.18	0.50	0.10	0.03	0.100
No.1 busheling	Easy	Very inexpensive	60	95.7	98.0	0.15	0.01	0.30	0.010	0.020	0.04	0.07	0.03	0.01	0.008
Ironmaking slag scrap	Difficult	Expensive	125	90.0	91.5	4.50	1.50	1.20	0.130	0.040	0.05	0.05	0.02	0.01	0.005
Briquetted iron borings	Reactive at turndown	Cheap	180	88.9	90.0	3.00	1.80	0.65	0.100	0.090	0.40	0.20	0.40	0.02	0.015

Continued Table 4.4

Type of scrap	Melting	Relative cost	Bulk density /cwt·m^{-3} (45.36kg·m^{-3})	Yield /%	w(Fe) /%	w(C) /%	w(Si) /%	w(Mn) /%	w(P) /%	w(S) /%	w(Cr) /%	w(Cu) /%	w(Ni) /%	w(Mo) /%	w(Sn) /%
Home pit	Fair	Cheap	75	83.2	83.0	0.05	0.10	0.50	0.020	0.025	0.06	0.04	0.08	0.01	0.005
Home crops and skulls	Fair to difficult	Cheap	125	92.5	94.0	0.10	0.10	0.45	0.020	0.014	0.06	0.04	0.08	0.01	0.005
Hot metal	—	Moderate	—	90.7	93.8	4.50	0.50	0.50	0.065	0.040	0.02	0.02	0.02	0.01	0.005

Normally, the lighter scrap is loaded in the front, and the heavier scrap in the rear end of the box. This causes the lighter scrap to land first in the furnace as the scrap box is tilted. It is preferable that the lighter scrap fall on the refractory lining first, before the heavier scrap, to minimize refractory damage. Also, since heavy scrap is more difficult to melt than light scrap, it is preferable that it sits on top so that it is closest to the area of oxygen jet impingement and hence melt faster.

Scrap pieces that are too large to be charged into the furnace are cut into smaller pieces by means of shears or flame cutting. Thin, small pieces of scrap such as sheet shearing and punching are compressed into block like bundles called bales using special hydraulic presses. Normally, larger, heavier pieces of scrap are more difficult to melt than lighter, smaller ones.

Unmelted scrap can cause significant problems in process control. It may result in high temperatures or missed chemistries at turndown. Bottom or mixed blowing, which can significantly enhance the mixing characteristics in the furnace, improves scrap melting of larger pieces.

Stable elements present in scrap, such as copper, molybdenum, tin and nickel cannot be oxidized and hence cannot be removed from metal. These elements can only be diluted. Detinned bundles, where tin is removed by shredding and treating with NaOH and then rebated, are available but at considerably higher cost. Elements such as aluminum, silicon and zirconium can be fully oxidized from scrap and become incorporated in the slag. Elements which fall in the middle category in terms of their tendency to react, such as phosphorus, manganese and chromium distribute themselves between the metal and slag. Zinc and lead are mostly removed from scrap the bath as vapor.

Most steelmaking shops typically use about 20% to 35% of their total metallic charge as scrap, with the exact amount depending on the capacity of the steelmaking process. Much of this capacity depends on factors like the silicon, carbon and temperature of the hot metal, use of a post combustion lance, and external fuels charged, such as anthracite coal. The scrap ratio is also influenced by the relative cost of scrap and hot metal.

4.3.4 High Metallic Alternative Feeds

Direct reduced iron (DRI) is used in some steelmaking shops as a coolant as well as a source of iron units. DRI typically contains about 88%-94% total iron (about 85%-95% metallization), 0.5%-3%C, 1%-5% SiO_2, 3%-8% FeO and small amounts of CaO, MgO and Al_2O_3. DRI may contain phosphorus in the range of 0.005%-0.09%, sulfur in the range 0.001%-0.03% and low concentrations of nitrogen (usually less than 20ppm).

DRI is normally fed into the BOF in briquetted form size at approximately. The DRI briquettes are passivated (by coating or binder) to eliminate any tendency to pyrophoricity (spontaneous burning) so that they can be handled conveniently in the steelmaking shop. DRI is usually fed into the steelmaking furnace through the bin system.

Certain elements such as nickel, copper and molybdenum can be added to the heat with the scrap charge. These elements do not oxidize to any significant level and they dissolve evenly in the metal during the oxygen blow. These additions can also be made after the oxygen blow, or in the ladle during tapping.

4.3.5 Oxide Additions

4.3.5.1 Iron Oxide Materials

Iron ore is usually charged into the BOF as a coolant and it is often used as a scrap substitute. Iron ores are available in the form of lumps or pellets, and their chemical compositions vary from different deposits as shown in Table 4.5. Iron ores are useful scrap substitutes as they contain lower amounts of residual elements such as copper, zinc, nickel, and molybdenum. The cooling effect of iron ore is about three times higher than scrap. The reduction of the iron oxide in the ore is endothermic and higher amounts of hot metal and lower amounts of scrap are required when ore is used for cooling. Iron ores must be charged early in the blow when the carbon content in the bath is high to effectively reduce the iron oxide. The reduction of the iron oxides in the ore produces significant amounts of gas, and consequently increases slag foaming and the tendency to slop. Late ore additions have a detrimental affect on iron yield and end point slag chemistry. If only ore is used as a coolant just before tap, the slag becomes highly oxidized and fluid, enhancing slag carryover into the ladle. The delay in the cooling reaction from the unreduced ore causes a sudden decrease in temperature or a violent ladle reaction resulting in over-oxidation of the steel.

Table 4.5 Iron ore chemical compositions   (%)

Ore	Country	$w(Fe)$	$w(SiO_2)$	$w(Al_2O_3)$	$w(CaO)$	$w(MgO)$	$w(P)$	$w(S)$	$w(Mn)$
Minnesota	USA	54.3	6.8	0.4	0.25	0.10	0.23	—	1.0
Carol lake	Canada	64.7	3.9	0.35	—	—	0.005	—	0.2
Cerro Bolivar	Venezuela	63.7	0.75	1.0	0.3	0.25	0.09	0.03	0.02
Goa	India	57.8	2.5	6.5	0.7	0.3	0.04	0.02	—
Itabira	Brazil	68.9	0.35	0.6	—	—	0.03	0.01	0.05
Tula	USSR	52.2	10.1	1.25	0.3	0.1	0.06	0.1	0.35

4.3.5.2 Waste Oxides

Economic and environmental issues have driven steel producers to recycle the waste iron oxides generated in the process. The increasing price of scrap, in addition to the increasing costs involved in the environmentally safe disposal of waste oxides, have encouraged steelmakers to recycle these materials back into the steelmaking process. Throughout the plant, various waste oxides and mill scales are collected and used in the sinter plant to produce some of the feed for the blast furnace. However, this does not consume all available oxides. In recent times, methods have been developed to substitute waste oxides in the BOF in place of ore. Mill scale has been used as a coolant in the BOF in amounts ranging from 5000 to 25,000lb(2268-11,400kg). Mill scale was found to be very effective in increasing the hot metal to scrap ratio; however, it causes heavy slopping during the process. Mill scale and other iron oxide additions are reduced during the main blow releasing iron and oxygen. This additional oxygen becomes available for carbon removal thus speeding up the overall reaction. Slopping is likely caused by the increased slag volume associated with using more hot metal (more pounds of Si and C generate more SiO_2 and CO, respectively) and by the increased reaction rate.

Waste oxide briquettes (WOB) containing steelmaking sludges, grit, and mill scale have also been charged into the furnace as a scrap substitute. The waste oxides collected from the BOF fumes during the blow are high in iron content, typically more than 60%. These fumes, fines (sludge) and coarse (grit) waste oxide particles are blended, dried, mixed with lime and binders, and pressed into pillow-shaped briquettes. The briquettes are then cured for over 48 hours to remove their moisture. A typical composition of WOBs is 35% sludge, 20% grit, and 45% mill scale, see Table 4.6.

Table 4.6 WOB chemical compositions

Composition	w/%
Total Fe	55-62
Metallic Fe	3-5
FeO	38-46
Fe_2O_3	29-32

Additions of WOBs are made early in the oxygen blow when the carbon content in the bath is high to ensure the reduction of ferrous, ferric, and manganese oxides to metallic iron and manganese. If the WOBs are added late in the blow, the oxides are likely to stay unreduced, resulting in yield loss, slopping, and a highly oxidized slag at turndown. WOBs are about two times better coolants than scrap, because their oxide reduction is endothermic, and therefore a higher hot metal ratio is required when WOBs are used for cooling—a situation similar to using ore. Various studies show that using WOBs causes no adverse effects on lining wear, molten iron yield, turndown performance and ladle slag FeO in the BOF.

4.3.6 Fluxes

4.3.6.1 Burnt Lime

In basic oxygen steelmaking, burnt lime consumption ranges from 40 to 100 lb(18.1-45.4kg) per net ton of steel produced. The amount consumed depends on the hot metal silicon, the proportion of hot metal to scrap, the initial (hot metal) and final (steel aim) sulfur and phosphorus contents. Burnt lime is produced by calcining limestone ($CaCO_3$) in rotary, shaft, or rotary hearth type kilns. The calcining reaction is given below:

$$CaCO_3 + Heat \longrightarrow CaO + CO_2 \qquad (4.1)$$

The calcination of high-calcium limestone will produce burnt lime containing about 96% CaO, 1% MgO, and 1% SiO_2. The sulfur content in burnt lime ranges from 0.03% to 0.1%. Most shops require less than 0.04% S in the lime to produce low sulfur steels. Since an enormous amount of burnt lime is charged into the BOF within a short period of time, careful selection of the lime quality is important to improve its dissolution in the slag. In general, small lump sizes [1/2-1 inch (1.27-2.54cm)] with high porosity have higher reactivity and promote rapid slag formation. The most common quality problems with either burnt or dolomitic lime are uncalcined inner cores, excess fines and too low a reactivity (calcined too hot or too long).

4.3.6.2 Dolomitic Lime

Dolomitic lime is charged with the burnt lime to saturate the slag with MgO, and reduce the dissolution of dolomite furnace refractories into the slag. Typically dolomitic lime contains about 36%-42% MgO and 55%-59% CaO. Similarly, the dolomitic stone contains about 40% $MgCO_3$. The dolomitic lime charge into the BOF ranges from 30 to 80 lb (13.6-36.3kg) per net ton of steel produced, which represents about 25%-50% of the total flux charge into the furnace (burnt plus dolomitic lime). The large variation in these additions strongly depends on experience and adjustments made by the steelmakers. These are based on observations of chemical attack of the slag on furnace refractories. Most of the dolomitic lime produced in the United States is obtained by calcining dolomitic stone in rotary kilns. The calcining reaction of the dolomitic stone is similar to that of limestone:

$$MgCO_3 + Heat \longrightarrow MgO + CO_2 \qquad (4.2)$$

In some BOF operations dolomitic stone is added directly into the furnace as a coolant, and as a source of MgO to saturate the slag. It can also be added to stiffen the slag prior to slag splashing.

It is important for the steelmaker to control the chemistry and size of the dolomitic lime.

4.3.6.3 Limestone

In most BOF shops limestone ($CaCO_3$) or dolomitic stone, ($CaCO_3 \cdot MgCO_3$) is frequently used as a coolant rather than as a flux. Limestone is commonly used to cool the bath if the turndown temperature is higher than the specified aim. When limestone is heated, the endothermic calcining reaction occurs producing CaO and CO_2, causing a temperature drop in the furnace. The extent of the temperature drop just before tap depends on the furnace size and slag conditions and is known for each shop. For example, in a 300 ton heat, 1000 lb (453.6kg) of limestone will drop the temperature of the bath by about 6℃ (10°F).

4.3.6.4 Fluorspar

Calcium fluoride or fluorspar (CaF_2) is a slag fluidizer that reduces the viscosity of the slag. When added to the BOF it promotes rapid lime (CaO) dissolution in the slag by dissolving the dicalcium silicate ($2CaO \cdot SiO_2$) layer formed around the lime particles which retards the dissolution of the lime in the slag. In recent times, fluorspar has been used very sparingly because of its very corrosive attack of all types of refractories, including both furnace and ladle. Also, the fluorides form strong acids in the waste gas collection system which corrode structural parts and are undesirable emissions.

4.3.7 Oxygen

In modern oxygen steelmaking processes a water-cooled lance is used to inject oxygen at very high velocities onto a molten bath to produce steel. With the increasing demands to produce higher quality steels with lower impurity levels, oxygen of very high purity must be supplied. Therefore, the oxygen for steelmaking must be at least 99.5% pure, and ideally 99.7%-99.8% pure. The remaining parts are 0.005%-0.01% nitrogen and the rest is argon.

In top-blown converters, the oxygen is jetted at supersonic velocities (Mach>1) with convergent divergent nozzles at the tip of the water-cooled lance. A forceful gas jet penetrates the slag and impinges onto the metal surface to refine the steel. Today, most BOFs operate with lance tips containing 4-5 nozzles and oxygen flow rates (in 230-300 ton converters), that range from 640-900Nm^3/min (22,500-31,500 scfm). Fig. 4.6 shows an schematic of a typical five nozzle lance tip. The tip is made of a high thermal conductivity cast copper alloy with precisely machined nozzles to achieve the desired jet parameters. The nozzles are angled about 12° to the centerline of the lance pipe and equally spaced around the tip. The tip is welded to a 12 inch (30.48cm) seamless steel pipe (lance barrel) about 60 feet (18.3m) long. Cooling water is essential in these lances to keep them from burning up in the furnace. At the top of the lance, armored rubber hoses are connected to a pressure-regulated oxygen source and to a supply of recirculated cooling water. Details of the convergent-divergent nozzles are also shown in Fig. 4.6. As the oxygen passes the converging section it is accelerated and reaches sonic velocity (Mach=1) in the cylindrical throat section. Then it expands in the diverging section and its temperature and pressure decreases while its velocity increases to supersonic levels (Mach > 1). The supersonic jets are at an angle of about 12° so that they do not interfere with each other.

In bottom-blown converters, the oxygen is injected through the bottom of the vessel using a series of tuyeres. About 14-22 tuyeres are used to blow about 4.0-4.5Nm^3 of oxygen per minute per ton of steel. Powdered lime, mixed with the oxygen, is usually injected through the liquid bath to improve lime dissolution and hence slag formation during the blow. The tuyeres consist of two concentric pipes, where oxygen flows through the center pipe and a hydrocarbon fluid, such as natural gas or propane, used as a coolant, flows through the annular space between both pipes.

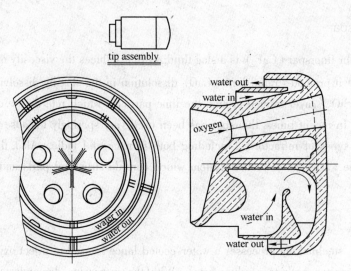

Fig. 4.6 Convergent-divergent nozzles

4.4 Process Reactions and Energy Balance

4.4.1 Reactions in BOF Steelmaking

In the oxygen steelmaking process, impurities such as carbon (C), silicon (Si), and manganese (Mn) dissolved in the hot metal are removed by oxidation to produce liquid steel. Hot metal and scrap are charged into the furnace and high-purity oxygen gas is injected at high flow rates, through a lance or tuyeres, to react with the metal bath. The oxygen injection process, known as the blow, lasts for about 16-25 minutes and the oxidation reactions result in the formation of CO, CO_2, SiO_2, MnO and iron oxides. Most of these oxides are dissolved with the fluxes added to the furnace, primarily lime (CaO), to form a liquid slag that is able to remove sulfur (S) and phosphorus (P) from the metal. The gaseous oxides, composed of about 90% CO and 10% CO_2, exit the furnace carrying small amounts of iron oxide and lime dust. Typical oxygen flow rates during the blow range between 2-3.5Nm^3 per minute per ton of steel (70-123 scfm per ton), and in general the rate of oxygen injection is limited either by the capacity of the hood and gas cleaning system or by the available oxygen pressure.

The commercial success of oxygen steelmaking is mainly due to two important characteristics. First, the process is autogenous meaning that no external heat sources are required. The oxidation reactions during the blow provide the energy necessary to melt the fluxes and scrap, and achieve the desired temperature of the steel product. Second, the process is capable of refining steel at high production rates. The fast reaction rates are due to the extremely large surface area available for reactions. When oxygen is injected onto the metal bath a tremendous amount of gas is evolved forming an emulsion with the liquid slag and with metal droplets sheared from the bath surface by the impingement of the oxygen jet. This gas metal-slag emulsion, shown in Fig. 4.7, generates

4.4 Process Reactions and Energy Balance

large surface areas that increase the rates of the refining reactions. Therefore, only a brief discussion of the sequence of these reactions during the blow is provided in this section.

4.4.1.1 Carbon Oxidation

Decarburization is the most extensive and important reaction during oxygen steelmaking. About 4.5% carbon in the hot metal is oxidized to CO and CO_2 during the oxygen blow, and steel with less than 0.1% carbon is produced. The change in the carbon content during the blow is illustrated in Fig. 4.8, which shows three distinct

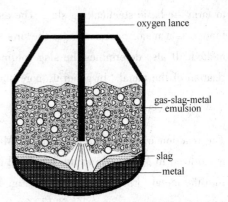

Fig. 4.7 Physical state of the BOF in the middle of the blow

stages. The first stage, occurring during the first few minutes of the blow, shows a slow decarburization rate as nearly all the oxygen supplied reacts with the silicon in the metal. The second stage, occurring at high carbon contents in the metal, shows a constant higher rate of decarburization and its controlled by the rate of supplied oxygen. Finally, the third stage occurs at carbon contents below about 0.3%, where the decarburization rate drops as carbon becomes less available to react with all the oxygen supplied. At this stage, the rate is controlled by mass transfer of carbon, and the oxygen will mostly react with iron to form iron oxide. Also in this stage, the generation of CO drops and the flame over the mouth of the furnace becomes less luminous, and practically disappears below about 0.1% carbon.

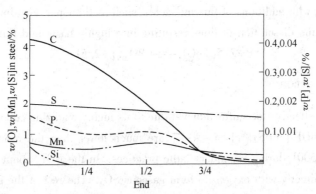

Fig. 4.8 Change in melt composition during the blow

4.4.1.2 Silicon Oxidation

The strong affinity of oxygen for Silicon will result in the removal of almost all the Si early in the blow. The Si dissolved in the hot metal (0.25%-1.3%) is oxidized to very low levels (<0.005%) in the first three to five minutes of the blow as shown in Fig. 4.8. The oxidation of Si to silica (SiO_2) is exothermic producing significant amounts of heat which raises the temperature of the bath. It also forms a silicate slag that reacts with the added lime (CaO) and dolomitic lime (MgO) to form the basic steelmaking slag. The amount of Si in the hot metal is very important since its oxidation is a major heat source to the process and it strongly affects the amount of scrap that can be melted. It also determines the slag volume and consequently affects the iron yield and dephosphorization of the metal. In general, more slag causes less yield but lower phosphorus.

4.4.1.3 Manganese Oxidation

The reaction involving the oxidation of Mn in steelmaking is complex. In top-blown processes Mn is oxidized to MnO early in the blow and after most of the silicon has been oxidized, the Mn reverts into the metal. Finally, as shown in Fig. 4.8, towards the end of the blow the Mn in the metal decreases as more oxygen is available for its oxidation. In bottom-blown processes, such as the OBM (Q-BOP), a similar pattern is found, but the residual Mn content of the steel is higher than for top-blown processes due to better stirring.

4.4.1.4 Phosphorus Oxidation

Dephosphorization is favored by the oxidizing conditions in the furnace. The dephosphorization reaction between liquid iron and slag can be expressed by reaction (4.3). Phosphorus removal is favored by low temperatures, high slag basicity (high CaO/SiO_2 ratio), high slag FeO, high slag fluidity, and good stirring. The change in the phosphorus content of the metal during blow is shown in Fig. 4.8. The phosphorus in the metal decreases at the beginning of the blow, then it reverts into the metal when the FeO is reduced during the main decarburization period, and finally decreases at the end of the blow. Stirring improves slag-metal mixing, which increases the rate of dephosphorization. Good stirring with additions of fluxing agents, such as fluorspar, also improves dephosphorization by increasing the dissolution of lime, resulting in a highly basic and fluid liquid slag.

$$P + 2.5(FeO) = (PO_{2.5}) + 2.5Fe \qquad (4.3)$$

4.4.1.5 Sulfur Reaction

The BOF is not very effective for sulfur removal due to its highly oxidizing conditions. Sulfur distribution ratios in the BOF $[w(S)/w[S] = 4\text{-}8]$ are much lower than the ratios in the steel ladle $[w(S)/w[S] = 300\text{-}500]$ during secondary ladle practices. In the BOF, about 10%-20% of sulfur in the metal reacts directly with oxygen to form gaseous SO_2. The rest of the sulfur is removed by the following slag-metal reaction:

$$S + (CaO) + Fe = (CaS) + (FeO) \qquad (4.4)$$

Sulfur removal by the slag is favored by high slag basicities (high CaO/SiO_2 ratio), and low FeO contents. The final sulfur content of steel is also affected by the sulfur contained in the furnace charge materials, such as hot metal and scrap. The sulfur content in the hot metal supplied from the blast furnace generally ranges from 0.020% to 0.040%, and if the hot metal is desulfurized before steelmaking the sulfur content in the hot metal can be as low as 0.002%. Heavy pieces of scrap containing high sulfur contents must be avoided if low sulfur alloys with less than 60ppm (0.006%) of sulfur are being produced. For example a slab crop of 2273kg (5000lb) containing

0.25% S [5.7kg (12.5lb) of sulfur] can increase the sulfur content of steel by about 15ppm (0.0015%) in a 300 ton BOF.

4.4.2 Slag Formation in BOF Steelmaking

Fluxes are charged into the furnace early in the blow and they dissolve with the developing oxides to form a liquid slag. The rate of dissolution of these fluxes strongly affects the slag-metal reactions occurring during the blow. Therefore, it is important to understand the evolution of slag during the blow. Several investigators have studied slag formation in oxygen steelmaking, and a detailed review of these investigations is given by Turkdogan, and Deo and Boom.

At the beginning of the blow, the tip of the oxygen lance is kept high above the bath surface, at about 3.5m (12ft), which results in the formation of an initial slag rich in SiO_2 and FeO. During this period large amounts of burnt lime and dolomitic lime are charged into the furnace. The lance is then lowered and the slag starts to foam at around one third of the blow due to the reduction of the FeO in the slag in conjunction with CO formation. The drop in the FeO content in the slag is shown in Fig. 4.9. Also, as the blow progresses, the CaO dissolves in the slag, and the active slag weight increases. Finally, after three quarters into the blow, the FeO content in the slag increases because of a decrease in the rate of decarburization. The resulting slag at turndown in top-blown converters (BOF or BOP) have typical ranges: 42%-55% CaO, 2%-8% MgO, 10%-30% FeO_T, 3%-8% MnO, 10%-25% SiO_2, 1%-5% P_2O_5, 1%-2% Al_2O_3, 0.1%-0.3% S. In the OBM (Q-BOP) or bottom-blown converter, the total FeO in the turndown slag is lower, ranging between 4%-22%.

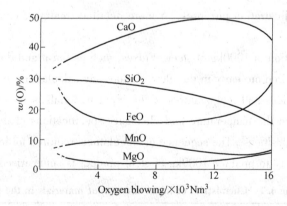

Fig. 4.9 Change in slag composition during the blow

During the blow, the temperature of the metal gradually increases from about 1350°C (2450°F) to 1650°C (3000°F) at turndown, and the slag temperature is about 50°C (120°F) higher than that of the metal. The slag at turndown may contain regions of undissolved lime mixed with the liquid slag, since the dissolution of lime is limited by the presence of dicalcium silicate ($2CaO \cdot SiO_2$) coating, which is solid at steelmaking temperatures and prevents rapid dissolution. The presence of MgO in the lime weakens the coating. Thus, charging MgO early speeds up slag forming due to quicker solution of lime.

4.4.3 Mass and Energy Balances

As shown in Fig. 4.10, hot metal, scrap, and iron ore are charged with the fluxes, such as burnt and dolomitic lime, into the furnace. Oxygen is injected at high flow rates and gases, such as CO and CO_2, and iron oxide fumes (Fe_2O_3) exit from the mouth of the furnace. At turndown, liquid steel and slag are the remaining products of the process. The oxidation reactions occurring during the blow produce more energy than required to simply raise the temperature of the hot metal, from about 1350℃ (2450°F) to the desired turndown temperature, and to melt the fluxes. Most of the excess heat is used to increase the amount of steel produced by melting cold scrap and by reducing iron ore to metal. Some heat is also lost by conduction, convection, and radiation to the surroundings.

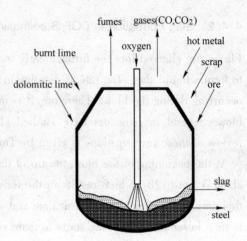

Fig. 4.10 Input and output materials in the BOF

It is important to exactly determine the amount of each material to charge and the amount of oxygen to blow to produce steel of desired temperature and chemistry. The specific method for determining these amounts varies with each BOF shop; however, in general these computations are based on mass and energy balance calculations. A simple mass and energy balance is presented in this text for illustration purposes. A more detailed treatment can be found in the literature.

Consider the production of 1000kg of steel. Fluxes, such as burnt and dolomitic lime, are added to the furnace with the iron ore early in the blow. Before any calculation can be made it is required to specify the compositions and temperatures of the input materials, such as hot metal, scrap, iron ore, and fluxes, and also the temperature and chemistry specifications of the steel product. Table 4.7 shows typical compositions. The sequence of calculations required to determine the amounts of input materials necessary to produce 1000kg of steel product is summarized as follows.

Table 4.7 Chemistries of input and output materials in the BOF

Element	Hot metal	Scrap	Steel
%Fe	93.61	99.493	99.797
%C	4.65	0.09	0.040
%Si	0.60	0.020	0.005
%Mn	0.45	0.360	0.138
%P	0.06	0.012	0.010
%S	0.01	0.025	0.010
Other	0.61		

Continued Table 4.7

Compound	Burnt lime	Dolo-lime	Slag	Iron ore
%CaO	96.0	58.0	47.86	—
%SiO$_2$	1.0	0.8	12.00	0.77
%MgO	1.0	40.5	6.30	—
%FeO$_T$	—	—	26.38	—
%Fe			—	65.8
%Al$_2$O$_3$	0.5	0.4	1.30	—
%MnO	—	—	5.00	0.20
%H$_2$O	1.3	1.4	—	—
%P$_2$O$_5$	—	—	1.16	—

4.4.3.1 Determination of the Flux Additions

The fluxes added to the process strongly depend on the hot metal silicon, the weight of hot metal, the lime to silica ratio (%CaO/%SiO$_2$), and the amount of MgO needed in the slag to avoid the wear of furnace refractories. The lime to silica ratio should range from two to four to achieve a basic slag during the blow. Also, approximately 6%-12% of MgO is required, depending on slag temperature and chemistry, to saturate the slag and consequently retard dissolution of the furnace refractories.

For a typical lime to silica ratio of four, each kg of SiO$_2$ in the slag requires 4kg of CaO. For the example shown in Table 4.8, about 11.26kg (24.78lb) of SiO$_2$ are produced from the oxidation of the hot metal silicon, and 47.92kg (105.42lb) of CaO per metric ton of steel are required to neutralize the SiO$_2$ in the slag. The amounts of burnt lime and dolomitic lime needed are computed from the CaO and MgO requirements as shown in Table 4.8. In actual BOF operations, higher dolomitic lime additions are made than those predicted by the present example to ensure MgO saturation.

Table 4.8 Material and energy balance for the production of 1000kg (2200lb) of steel

	Input			Output	
Hot metal	kg	lb	Steel	kg	lb
Fe	821.60	1807.5	Fe	997.97	2195.5
C	40.81	89.78	C	0.40	0.88
Si	5.26	11.58	Si	0.05	0.11
Mn	3.95	8.69	Mn	1.38	3.04
P	0.53	1.16	P	0.10	0.22
S	0.09	0.19	S	0.10	0.22
BF slag	5.40	11.88			
Total	877.64	1930.8	Total	1000.0	2200.0

Continued Table 4.8

Input			Output		
Scrap	kg	lb	slag	kg	lb
Fe	200.53	441.16	CaO	47.92	105.42
C	0.18	0.40	SiO_2	12.01	26.42
Si	0.04	0.09	MgO	6.35	13.97
Mn	0.72	1.60	FeO_T	26.42	58.12
P	0.02	0.05	MnO	5.00	11.01
S	0.05	0.11	P_2O_5	1.13	2.49
			Al_2O_3	1.3	2.86
Total	201.55	443.41	Total	100.13	220.29
Ore	kg	lb	Fumes	kg	lb
Fe	11.52	25.35	Fe	16.02	35.24
SiO_2	0.130	0.285			
MnO	0.034	0.075			
Other	5.134	11.294			
Total	16.818	37.0	Total	27.52	60.61
Fluxes	kg	lb	Gases	kg	lb
CaO	47.92	105.42	CO (90%)	85.26	187.57
MgO	6.35	13.97	CO_2 (10%)	14.89	32.75
SiO_2	0.53	1.17	Other	1.65	3.62
Total burnt time	40.78	89.70			
Total dolo-lime	15.67	34.48			
Oxygen gas	52.8Nm^3	1849.7scf			
	at 298K	at 21℃(70°F)			
Total O_2	75.34	165.75			
Total inputs	1227.8	2701.2	Total outputs	1227.8	2701.2

4.4.3.2 Determination of Oxygen Requirements

The volume of oxygen gas blown into the converter must be sufficient to oxidize the C, Si, Mn, and P during the blow, and it is computed from an oxygen balance as shown below. For the present example the oxygen required during the blow is about 52.8Nm^3 (1849.7scf) per metric ton of steel produced.

$$\begin{bmatrix} \text{Oxygen} \\ \text{injected} \end{bmatrix} = \begin{bmatrix} \text{Oxygen for the} \\ \text{oxidation reactions} \end{bmatrix} - \begin{bmatrix} \text{Oxygen supplied} \\ \text{by iron ore} \end{bmatrix} - \begin{bmatrix} \text{Oxygen dissolved} \\ \text{in steel at turndown} \end{bmatrix} \quad (4.5)$$

4.4.3.3 Determination of the Weight of Iron-Bearing Materials

In general four distinct iron-bearing materials are involved in oxygen steelmaking: hot metal, scrap, iron ore, and the steel product. Slag and fume are usually considered heat and iron losses.

The simultaneous solution of an iron mass balance and an energy balance permits the determination of the weights of two of the iron-bearing materials with a knowledge of the weights of the other two. For the example here, the product weight (1000kg), and the weight of the iron ore (16.8kg) are assumed to be known. Then the weights of the hot metal and scrap are computed to be 877.64kg (1930.8lb) and 201.55kg (443.41lb) respectively from the mass and energy balance shown below:

Mass balance for iron: (Iron input = Iron output)

Iron input = [Weight of Fe in hot metal] + [Weight of Fe in scrap] + [Weight of Fe in iron ore]

Iron output = [Weight of Fe in steel] + [Weight of Fe in slag] + [Weight of Fe in fumes]   (4.6)

Heat balance: (Heat input = Heat output)

Heat input = [Heat content in the hot metal] + [Heat of reactions] + [Heat of slag formation]

Heat output = [Sensible heat of steel] + [Sensible heat of slag] + [Sensible heat in gas and fume] + [Heat losses]   (4.7)

The heat added to the process comes from the heat content or enthalpy in the hot metal charged into the furnace at about 1343°C (2450°F), the heats of oxidation of elements, such as Fe, C, Si, Mn, P, and S, whose enthalpies are shown in Table 4.9, and the heats of formation of the different compounds in the slag.

Table 4.9 Enthalpies or heats of reactions

Oxidation reactions	Kilojoule	Heats of reaction		
		per mole of	BTU	per lb of
$C + 1/2O_2 = CO$	4173	C	3952	C
$C + O_2 = CO_2$	14,884	C	14,096	C
$CO + 1/2O_2 = CO_2$	4593	CO	4350	CO
$Si + O_2 = SiO_2$	13,927	Si	13,190	Si
$Fe + 1/2O_2 = FeO$	2198	Fe	2082	Fe
$Mn + 1/2O_2 = MnO$	3326	Mn	3150	Mn

These sources will provide the heat necessary to raise temperature of the steel and slag to the aim turndown temperature, and also to heat up the gases and fumes leaving the furnace. Furthermore, there is enough energy generated to overcome the heat losses during the process, to heat and melt coolants such as scrap and iron ore, and to reduce the iron oxide in the ore.

4.4.3.4 Determination of the Gases and Fumes Produced

The amounts of CO and CO_2 produced from decarburization are determined from a mass balance for carbon. The carbon removed from the bath is converted to approximately 90% CO and 10% CO_2.

With the gases about 1%-1.5% of iron is lost in the form of iron oxide fumes that exit from the mouth of the furnace.

4.4.3.5 Determination of the FeO in the Slag

The FeO in the slag is generally determined from empirical correlations, developed by each shop, between the slag FeO and the aim carbon and the lime to silica ratio. Other parameters are generally of much lower significance. This empirical relationship is one of the larger error sources in a material and energy balance algorithm, arising from analytical errors of iron oxides and slag sample preparation problems.

4.4.4 Tapping Practices and Ladle Additions

When the blow is completed, the lance is removed from the furnace and the vessel is rotated to a horizontal position towards the charging side for sampling. A steel sample is withdrawn from the bath for chemical analysis, and an expendable immersion-type thermocouple is used to measure the temperature of the melt. The steel sample is analyzed with a mass spectrometer, and the concentrations of the elements present in the steel are determined in approximately 3-5 minutes. If the steel is too hot, meaning that the measured temperature is higher than the aim temperature, it can be cooled by rocking the vessel, or by adding coolants such as iron ore or limestone. If the steel is too cold, or if the measured concentrations of elements such as carbon, phosphorus, and sulfur are higher than the aim concentrations specified, additional oxygen is blown into the furnace (reblow) for approximately 1-3 minutes. Once the heat meets the temperature and chemistry requirements, the furnace is rotated towards the taphole side and the steel is tapped or poured into a ladle.

Tapping a 300t heat takes from four to seven minutes and the time strongly depends on the conditions or diameter of the taphole. A good tapping practice is necessary to maximize yield, or the amount of steel poured into the ladle. Slag carryover from the BOF into the ladle must be minimized. Furnace slag contains high FeO, which reduces desulfurization in the ladle, and enhances the formation of alumina inclusions. Also, the P_2O_5 present in BOF slags is a source of phosphorus carried into the ladle. Therefore, over the years, extensive work has been done to develop slag free tapping techniques, and the most commonly used are described here. Good tap hole maintenance, combined with the ability of the operator to rotate the furnace quick enough when all the steel has been tapped, will reduce the amount of slag carryover into the ladle. Slag free tapping devices are now commonly used to help the operators reduce slag carryover. Different types of taphole plugs, such as balls and darts, shown in Fig. 4.11, are dropped into the furnace at tap. These Devices float at the slag-metal interface, and plug the taphole when the steel has emptied but before the slag can exit the furnace. There has been much debate over the effectiveness of these devices. Electromagnetic slag detection sensors installed around the taphole will detect the presence of slag in the stream and send a signal to alarm the furnace operator. One of the problems with these devices is that they can give false alarms from slag entrainment within the vortex of the steel stream and they require maintaining a taper in the taphole to work well.

With current steelmaking alloying practices, most of the alloys are added to the ladle. However, large amounts of non-oxidizable alloys such as nickel, molybdenum and copper are usually charged with the scrap as they resist oxidation during the blow. This practice will prevent big temperature drops in the ladle. In aluminum killed steels aluminum (Al) is used to deoxidize the steel and reduce the dissolved oxygen from approximately 500-1000ppm to less than 5ppm, and is generally the first addition made into the ladle during tap. For example, a heat with a Celox reading of 750ppm of dissolved oxygen requires approximately 365kg (800lb) of aluminum at tap. In semi-killed steels, deoxidation is done with ferrosilicon, and the dissolved oxygen in steel is only lowered to about 50ppm. Burnt lime is also added with the aluminum to satisfy

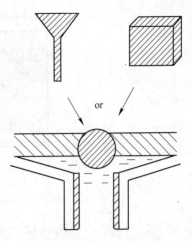

Fig. 4.11 Schematic of taphole plugs for BOF steelmaking

a lime to alumina (CaO/Al_2O_3) ratio of 0.8 to 1.2. This produces a liquid slag over the molten metal that thermally insulates the melt to avoid excess temperature losses, protects the melt from reoxidation from air, desulfurizes the steel, and removes alumina inclusions from the melt.

Ferromanganese is added via chutes located over the ladle, in large quantities, after the steel has been deoxidized by aluminum or silicon. The general rule of thumb is that the aluminum is added when the melt reaches approximately 1/3 of the ladle's height, and all the alloys should be added by the time the melt reaches 2/3 of the full ladle height. Slag modifiers, containing about 50% Al, are added to the slag near the end of tap to reduce the FeO content in the ladle slag originating from furnace slag carryover.

4.5 Process Variations

4.5.1 The Bottom-Blown Oxygen Steelmaking or OBM (Q-BOP) Process

The successful development and application of the shrouded oxygen tuyere in the late 1960s led to the development of the OBM (Q-BOP) process in the early 1970s. Oxygen in this process is injected into the bath through tuyeres inserted in the bottom of the furnace. Each tuyere is made from two concentric tubes forming an inner nozzle and an outer annulus. Oxygen and powdered lime are injected through the central portion of the tuyeres, while a hydrocarbon gas, typically natural gas or propane, is injected through the annular section between the two concentric pipes, as shown in Fig. 4.12. The endothermic decomposition of the hydrocarbon gas and the sensible heat required to bring the products of the decomposition up to steelmaking temperatures result in localized cooling at the tip of the tuyere. The localized cooling is enough to chill the liquid metal and form a porous mushroom on the tip of the tuyere and part of the surrounding refractory. This mushroom reduces the burn back rate of the tuyere, and the wear of the surrounding refractory. The injected lime pro-

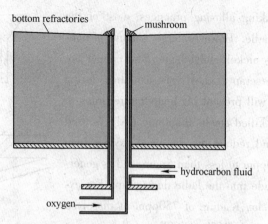

Fig. 4.12 Schematic drawing of an OBM (Q-BOP) tuyere

vides additional cooling to the tuyere, and results in better slag refining characteristics.

Top lances in OBM (Q-BOP) furnaces have also been adopted, mainly for the purposes of increasing the post-combustion of the offgases within the furnace, and to control the buildup of slag and metal in the furnace cone area. Top lances used in OBM (Q-BOP) furnaces are normally stationary, since they are not used for refining purposes. Tuyeres, located in the upper cone area of furnaces with a heat size larger than approximately 150 tonnes have also been used, but typically result in higher refractory wear. For this reason, their application has been limited to shops which require increased scrap melting capabilities (resulting in shorter lining lives), and with a heat size smaller than 150 tonnes.

4.5.1.1 Plant Equipment

The injection of oxygen through the bottom in the OBM (Q-BOP) process, with a fraction of the total oxygen through a stationary top lance, results in the need for a low building, and consequently in lower greenfield construction costs. Oxygen, a hydrocarbon fluid (natural gas or propane), nitrogen and argon are gases used in the OBM (Q-BOP) process. These gases are metered and controlled, and introduced through rotary joints located in the trunnion pins. The oxygen, aside from being used as the main process gas, is also used as the transport gas for the pulverized burnt lime. A high-pressure injector contains the burnt lime, which is transported by the oxygen through one of the trunnion pins into a lime and oxygen distributor, and then to the individual tuyeres in the furnace bottom. The hydrocarbon fluid is transported to the bottom of the furnace through the opposite trunnion, to avoid the possibility of leakage and of its mixing with oxygen in the transport line. The hydrocarbon fluid is then distributed to each tuyere. In some instances, the flow of the hydrocarbon fluid is controlled individually for each tuyere prior to entering the rotary joint. Nitrogen is used to protect the tuyeres from plugging during furnace rotation. It can also be used to increase the nitrogen content in the steel. Argon can be used to minimize the nitrogen pickup in the steel, and to produce lower carbon steels than in the BOF process, without excessive yield losses, and with low FeO contents in the slag.

Gas flow through the tuyeres has to be maintained above sonic flow to prevent penetration of steel into the tuyeres, and subsequent plugging. For this reason, the sonic flow is maintained during rotation of the furnace for turndown and for tapping of the steel. The ejections of metal and slag resulting from the injection of this high flow through the tuyeres during rotation, require the complete enclosure of the OBM (Q-BOP) furnaces. Movable doors are used in the charge side for this purpose. The complete enclosure also results in lower fugitive emissions during turndown and tapping than in the BOF furnaces, which usually are not enclosed.

Typically 12-18 tuyeres are used in the bottom, depending on the furnace capacity. The tuyeres are located in the refractory bottom on two rows that run from one trunnion to the other. The locations of these two rows are selected so that they are above the slag line during turndown and tapping to allow the reduction of the gas flow during these periods. The inside of the inner pipe of the tuyere is typically lined with a mullite sleeve to prevent excessive wear of the pipe by the burnt lime. The inside diameter of the ceramic liner is 1-1.5 inches (2.54-3.81cm). Stainless steel is normally used for the inner pipe, and carbon steel for the outer pipe, although copper has also been used.

The higher refractory wear observed in the vicinity of the tuyeres, due to the high temperature gradients experienced by the refractories during a heat cycle and the high temperature generated around them, results in bottom lives of 800-2500 heats, depending on tap temperature, turndown carbon content, etc. Since barrel lives approach 4000-6000 heats, the furnace is designed to include a replaceable bottom. When the bottom thickness is too thin for safe operation, it is removed and a bottom with new refractories and tuyeres is installed. The furnace can be back in operation with a new bottom in less than 24 hours.

4.5.1.2 Raw Materials

A distinct advantage of the OBM (Q-BOP) process is its capability to melt bigger and thicker pieces of scrap than the BOF process. Sections with thicknesses of up to two feet are melted routinely. This expands significantly the types of scrap that can be used, and lowers their preparation costs. There is no unmelted scrap at the end of the blow in the OBM (Q-BOP) process.

The burnt lime used for slag formation is pulverized and screened to less than 0.1 millimeters. It is sometimes treated to improve its flowability during pneumatic transport with oxygen. The dolomitic lime in the OBM (Q-BOP) process is essentially similar to that used in the BOF process, and is charged through top bins, if available. In shops where overhead bins are not available, a pulverized blend of burnt lime and dolomitic lime is injected through the bottom tuyeres to achieve the desired MgO content in the slag. The rest of the raw materials are the same as for the BOF process.

4.5.1.3 Sequence of Operations

After the steel is tapped, the furnace is rotated to the vertical position, and nitrogen is blown to splash the slag onto the furnace walls. This results in a coating that extends the life of the furnace barrel. The furnace is also rocked to coat the bottom with slag. This operation can be done with the slag as is, or with conditioned slag. The furnace is then ready to receive the scrap and hot metal. Nitrogen is injected at sonic flow to protect the tuyeres during the hot metal charge. The furnace is rotated to its vertical position, and the bottom and top oxygen blow are started. Burnt lime is injected with the oxygen through the bottom, and the dolomitic lime is added through the top at the beginning of the blow. Typically the lime is injected within the first half of the oxygen blow. When the calculated oxygen amount has been injected, the gas is switched to nitrogen or argon and the furnace is rotated for sampling. A sample for chemical analysis is taken, and the temperature and oxygen activity are measured. If the desired temperature and chemistry have been obtained, the heat can then be tapped. If necessary, small adjustments in temperature and chemistry can be made by injecting additional oxygen through the bottom, by injecting more lime, or by cooling the heat with ore or raw dolomite. Since the process is very reproducible, the heat is normally tapped after these adjustments, without making another temperature measurement or taking another sample for chemical analysis. Shops so equipped can take a sample for chemical analysis, temperature and carbon content measurement with sub lances a few minutes before the end of the oxygen blow. Any necessary adjustments can be made during the blow to obtain the aim steel temperature and chemistry. The turndown step can thus be avoided, decreasing the tap-to-tap time, and increasing the productivity of the shop.

4.5.1.4 Process Characteristics

The injection of the oxygen and hydrocarbons through bottom tuyeres results in distinct process characteristics. The oxygen reacts directly with the carbon and silicon in the liquid iron melt, resulting in lower oxidation levels in the metal and slag at the end of the blow. The bottom injection also results in very strong bath mixing. Steel decarburization is enhanced by the strong bath agitation, particularly during the last portion of the blow, when mass transfer of the carbon in the melt controls the rate of decarburization at carbon contents below 0.3%. This results in less iron being oxidized and lost to the slag, as shown in Fig. 4.13, and in less dissolved oxygen in the steel at turndown, as shown in Fig. 4.14. The manganese content at turndown is also higher than in the

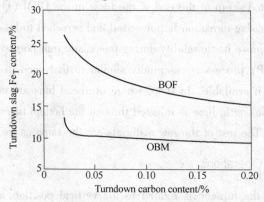

Fig. 4.13 Slag Fe_T content as a function of the carbon content at turndown in the U.S. Steel Gary Works BOF and OBM (Q-BOP) shops

top-blown vessels, as shown in Fig. 4.15, due to the lower bath and slag oxidation. The variability in blow behavior introduced by the top lance in top-blown vessels is eliminated. By injecting the oxygen and lime through the bottom tuyeres in a controlled manner, a highly reproducible process control is obtained.

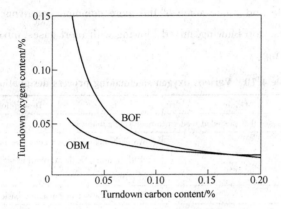

Fig. 4.14 Dissolved oxygen content as a function of carbon content at turndown in the U.S. Steel Gary Works BOF and OBM (Q-BOP) shops

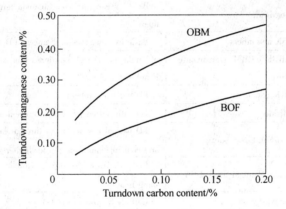

Fig. 4.15 Manganese content in the steel as a function of carbon content at turndown in the U.S. Steel Gary Works BOF and OBM (Q-Bop) shops

4.5.1.5 Product Characteristics

In general, the same steel grades that can be produced in top-blown vessels can be produced by the OBM (Q-BOP). Additionally, the better mixing attained in the bath allows the production of steels with carbon contents of 0.015%-0.020% without excessive bath and slag oxidation and without vacuum decarburization. The slightly higher hydrogen contents obtained at turndown can be lowered by flushing the bath with nitrogen, argon, or a mixture of argon and oxygen at the end of the blow. For the production of low-nitrogen steels, argon can be used to minimize the nitrogen pickup during rotation, or by flushing the bath at the end of the blow. Nitrogen can also be alloyed into the steel by purging the melt with nitrogen at the end of the oxygen blow.

4.5.2 Mixed-Blowing Processes

Virtually every company has invented or modified a version of the oxygen steelmaking process to suit its own situation. Consequently, there are many designations for fairly similar processes. Table 4.10 attempts to translate and relate some of the more common acronyms. This table is divided into four broad categories: top blowing, mixed blowing with inert gases, mixed blowing with bottom oxygen, and bottom blowing.

Table 4.10 Various oxygen steelmaking process designations

Process name	Origin	Description
Obsolete bottom-blown processes that preceded modern furnace configurations:		
Bessemer converter		An early bottom-blown converter developed by Henry Bessemer in 19th Century England, blowing air through simple tuyeres using acid refractories
Thomas converter		Similar to bessemer but using basic refractories in US
All top blowing:		
LD	Voest, Austria	Linz-donawitz. First top-blown process with water-cooled lance, lump lime
BOF	Worldwide	Basic oxygen furnace. Common term for LD top lance-blown, lump lime
BOP	USX and others	Basic oxygen process. Same as LD and BOF
LD-AC	ARBED/CRM Luxembourg; France	Similar to LD with powdered lime added through lance, for high P hot metal
LD-CL	NKK, Japan	LD but lance rotates
LD-PJ	Italsider	LD with pulsed jets, not in current use
ALCI	ARBED, Luxembourg	LD basically with Ar/N_2 through lance. Post combustion ports and coal injection from top for higher scrap melting
LD-GTL	Linde/National Steel, USA	LD with Ar or N_2 through top lance to limit over-oxidation, lump lime
AOB	Inland, Union	Carbide Similar to LD-GTL
Z-BOP	ZapSib Russia	Basically LD or BOF. Various process methods of adding additional fuels to increase scrap melting. Some include preheat cycles. Can melt 100% scrap
Mixed blowing, inert stirring gases:		
LBE	ARBED, Luxembourg IRSID, France	Lance bubbling equilibrium. LD with permeable plugs on bottom for inert gas. Lump lime
LD-KG	Kawasaki, Japan	LD with small bottom tuyeres Ar and/or N_2
LD-KGC	Kawasaki, Japan	LD with number of small nozzles using Ar, N_2, CO for inert gas bottom stirring. Unique in that it uses recycled CO as a stirring gas. Lump lime
LD-OTB	Kobe, Japan	Similar to LD-KG
LD-AB	Nippon Steel, Japan	LD with simple tuyeres to inject inert gas. Lump lime
NK-CB	NKK, Japan	Top-blown LD with simple bottom tuyere or porous plugs to introduce $Ar/CO_2/N_2$, lump lime

4.5 Process Variations

Continued Table 4.10

Process name	Origin	Description
Mixed blowing with oxygen and/or inert bottom gases:		
OBM-S	Maxhutte, Germany Klockner, Germany	Mostly bottom OBM type with top oxygen through natural gas shrouded side tuyere, powdered lime through bottom
K-BOP	Kawasaki Japan	Top and bottom blowing. Natural gas shrouded bottom tuyeres, powdered lime through tuyeres
TBM	Thyssen, Germany	Top and bottom stirring with bottom nozzles and N_2/Ar
LET	Solmer, France	Lance Equilibrium Tuyeres. Top blowing with 15%-35% bottom blown with fuel oil shrouded tuyeres
LD-OB	Nippon Steel, Japan	OBM tuyeres (natural gas shrouded) on bottom with top lance, lump lime
STB	Sumitomo, Japan	Mostly top-blown with lance, with special tuyere on bottom. Inner pipe O_2/CO_2, Outer pipe CO_2/N_2/Ar. Lump lime
STB-P	Sumitomo, Japan	Similar to STB except powdered lime through top lance for phosphorus control
All bottom blowing:		
OBM	Maxhutte, Germany	Original 100% bottom-blown. Natural gas shrouded tuyeres, powdered lime through tuyeres
Q-BOP	USX, USA	OBM type 100% bottom-blown. Natural gas shrouded tuyeres, powdered lime through tuyeres
KMS	Klockner, Germany	Similar to OBM. Early trials of oil shrouded tuyeres, now use natural gas. Can inject powdered coal from bottom for more scrap melting
KS	Klockner, Germany	Similar to KMS only modified for 100% scrap melting

Note: Bold lettered processes are still in regular use today.

The BOF is the overwhelmingly popular process selection for oxygen steelmaking. The usual reason to modify the BOF configuration is to lower the operating cost through better stirring action in the steel bath. It has been found by a number of companies and investigators that additional stirring of the bath from the furnace bottom reduces FeO in the slag. Lower FeO results in higher yield, and fewer oxidation losses of metallics. Bottom stirring increases slag forming, particularly if powdered limes are injected into the bath.

4.5.2.1 Bottom Stirring Practices

Inhomogenieties in chemical composition and temperature are created in the melt during the oxygen blow in the top-blown BOF process due to lack of proper mixing in the metal bath. There is a relatively dead zone directly underneath the jet cavity in the BOF.

Bottom stirring practices using inert gases such as nitrogen and argon are being used extensively to improve the mixing conditions in the BOF. The inert gases are introduced at the bottom of the furnace by means of permeable elements (LBE process) or tuyeres. In a typical practice, nitrogen gas is introduced through tuyeres or permeable elements in the first 60%-80% of the oxygen blow, and argon gas is switched on in the last 40%-20% of the blow. The rapid evolution of CO in the first part of the oxygen blow prevents nitrogen pickup in the steel.

Some of the effects of bottom stirring and the resulting improved mixing include:

1. Decreased FeO content in slag. Better mixing conditions in the vessel causes the FeO in the slag to be closer to equilibrium conditions, which results in lower concentrations of FeO in the slag. Plant studies have shown that for low carbon heats, bottom stirring can cause a reduction in the FeO level in slag by approximately 5%. This results in better metallic yield, lower FeO level in the ladle slag and reduced slag attack on the refractories. Improvements in iron yield by as much as 1.5% or more have been reported. Lower levels of FeO in the steelmaking slag reduces the amount of heat generated during the oxygen blow, and hence reduces the maximum amount of scrap that can be charged in a heat.

2. Reduced dissolved oxygen in metal. A study shows that bottom stirring can reduce the dissolved oxygen level in a low carbon heat by approximately 225ppm. This lowering of dissolved oxygen leads to lower aluminum consumption in the ladle. Studies have shown aluminum savings of about 0.3lb/(0.14kg/t) due to bottom stirring.

3. Higher manganese content in the metal at turndown. An increase of approximately 0.03% in the turndown manganese content of the metal has been shown. This leads to a reduction in the consumption of ferro-manganese.

4. Sulfur and Phosphorus removal. Bottom stirring has been found to enhance desulfurization due to improved stirring. However, phosphorus removal has not been found to improve substantially in some studies. Although bottom stirring drives the dephosphorization reaction towards equilibrium, the reduced levels of FeO in steelmaking slags tend to decrease the equilibrium phosphorus partition ratio $\{w(P)/w[P]\}$.

4.5.2.2 Bottom Stirring Maintenance Problems

The price for the bottom stirring benefits is in maintaining a more complicated process. This complexity varies depending on the type of bottom stirring devices installed. Generally, the simpler the device is to maintain, the lesser is the benefit obtained.

Starting at the simple end, porous plugs or permeable elements do not require maintaining gas pressure when the stirring action is not required. While gas under pressure can permeate these devices, steel does not penetrate, even when the gas is turned off. The disadvantage of the porous plugs is that they are not very effective stirrers. Two factors influence their poor performance. First, only a relatively small amount of gas can be introduced per plug. More gas flow requires more plugs thus adding complexity. Secondly, if the furnace bottom tends to build up, the plugs are covered over with a lime/slag mush and gas does not stir the steel bath. Rather, it escapes between the bricks and the protective mush. Another problem is at higher flow rates, the originally installed plugs do not last very long (1500-3000 heats) and replacements are shorter lived (200-2000 heats). The reasons for the shorter replacement plug life is not clear.

The use of the bottom stirring tuyeres (rather than plugs) can lead to stirring reliability and effectiveness problems. The tuyeres are known to get blocked early in the lining campaign when there is a buildup at the bottom of the furnace. The tuyere blockage can be prevented by maintaining proper gas flow rates through the tuyeres at different stages of the heat, by using high quality refractory bricks in the area surrounding the stirring elements or tuyeres, and by properly maintaining the bottom of the furnace by minimizing buildup.

4.5.2.3 Bottom Plug/Nozzle Configurations

There are basically three types of tuyeres used for mixed blowing. First, there is a refractory element that behaves much like porous plugs. This unit is made of compacted bricks with small slits. Like most tuyeres, it needs sufficient gas pressure to prevent steel penetration. This unit is more penetrating than porous plugs. Second, an uncooled tuyere is used to introduce large amounts of inert gases per nozzle. This results in local heavy stirring, which can more easily penetrate the buildup. Air or oxygen cannot be used because there is no coolant and the heat generated would make tuyere life too short to be practical. The third type is a fully cooled tuyere similar to the OBM (Q-BOP). Here, either inert gas or oxygen can be blown, causing very strong stirring and almost no problems penetrating bottom buildup.

In all cases the gas piping is routed through the furnace trunnions using rotary joints or seals to allow full rotation of the furnace.

4.5.3 Oxygen Steelmaking Practice Variations

4.5.3.1 Post-Combustion Lance Practices

Post combustion involves burning the CO gas produced by the reaction of carbon in the metal with jet oxygen, to CO_2 within the furnace. The heat generated as a result of this oxidation reaction can be used for any of the following purposes: to increase the scrap-to-hot metal ratio, to minimize the formation of skulls formed at the mouth of the vessel, to minimize the formation of lance skulls and to reduce the formation of "kidney" skulls in the upper cone region inside the BOF.

In a normal BOF operation, the percentage of CO in the off gas is approximately 80%-90% during the middle portion of the blow, when the rate of decarburization is maximum (Fig. 4.16). The oxidation of the CO gas to CO_2 generates 12 Mcal/Nm^3 CO (13.95kW · h/Nm^3, 1260 BTU/scf CO), which can melt more scrap and/or reduce skulling problems.

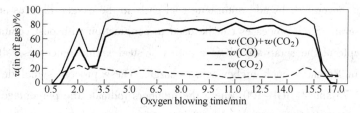

Fig. 4.16 The percentage of CO and CO_2 gases in the BOF off gas

Post combustion lances have primary nozzles as well as secondary nozzles which typically are located at about nine feet from the tip of the lance, (Fig. 4.17). The primary nozzles in PC lances are identical to those of the standard lances. They consist of four or five nozzles at the tip of the lance with inclination angles varying from 10° to 14°. Usually, there are eight secondary nozzles above the primary ones, directed out, at an angle of about 35° from the vertical. The oxygen flow rate through the secondary nozzles is considerably less than through the primary ones, usually 1.5%-15% of the total oxygen flow.

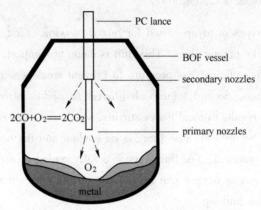

Fig. 4.17 Schematic of a post combustion lance in a topblown BOF

Studies show that post combustion has been successfully used to melt additional scrap in the BOF (achieving scrap-to-hot metal ratios of 25% or more), and to reduce lance and furnace mouth skulling problems. It has been found to increase the turndown temperature, if additional scrap or coolant is not added, to compensate for the extra heat generated. No effect on slag FeO and turndown manganese in metal has been found. There has been no effect on refractories when postcombustion is used at low flow rates in the skull control mode. However, if post-combustion is used in high flow, scrap melting mode, some care in design and operating parameters must be observed to avoid damaging the cone.

4.5.3.2 High Scrap Melting Practices

The ability to vary the scrap-to-hot metal ratio in a BOF is important because of the variations in scrap prices and the possibility of having hot metal shortages from time to time due to problems in the blast furnace. Several shops add anthracite coal early in the oxygen blow as an additional fuel to melt more of scrap. Such processes can be used in conjunction with post combustion lances.

A recent example of the scrap stretch is a series of steelmaking processes called Z-BOP that were developed in Russia to melt higher scrap charge. These processes include Z-BOP-30, Z-BOP-50, Z-BOP-75 and Z-BOP-100, where the numbers indicate the percentage of scrap charge in the BOP. The Z-BOP-100 process utilizes 100% scrap charge with no hot metal, using the conventional BOP with virtually no equipment modifications.

In the Z-BOP processes, the additional energy required to melt scrap comes from lump coal (anthracite and high volatile bituminous), which can be fed using a bin system. In the Z-BOP-100 process, the first batch of scrap is fed into the furnace using a scrap box, and coal is added in small batches while blowing oxygen. This process continues with additional scrap being charged. The number of scrap boxes charged can vary from three to five. The main blow commences when all the scrap is charged in the furnace, and small additions of coal are made throughout the main blow.

The typical tap-to-tap time for the Z-BOP-100 process is 64-72 minutes. No one has used these high scrap percent processes on a sustained basis because of damage to the furnace linings. Typical iron yield of the Z-BOP 100 process is about 80%, (91%-92% is normal for the BOF) and the slag FeO content is very high (45%-72%; normal for the BOF is approximately 25%). The sulfur content of tapped steel was not found to be higher than that of the scrap itself.

The major disadvantages of this process include decreased yield, increased slag FeO which can adversely affect the refractory lining, presence of tramp elements similar to EAF, and presence of fugitive emissions (SO_2 and NO_x). This process may be used during hot metal shortages, but is not proven for continuous use. Some shops have used a 70% or 80% scrap charge to accommodate a blast furnace reline and stretch a limited amount of hot metal from the remaining blast furnace(s).

4.5.3.3 Slag Splashing Refractory Maintenance

Any improvement in the furnace lining life is a boon for steelmakers due to the increased furnace availability and reduced refractory costs. Slag splashing is a technology which uses high pressure nitrogen through the oxygen lance after tapping the heat, and splashes the remaining slag onto the refractory lining (Fig. 4.18). The slag coating thus formed cools and solidifies on the existing refractory, and serves as the consumable refractory coating in the next heat.

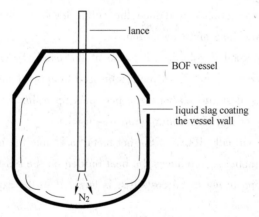

Fig. 4.18 Schematic of slag splashing in the BOF

Slag splashing is currently being used by several BOF shops, and has extended refractory lining life to record levels. A typical slag splash operation following the tapping of a heat is as follows:

1. The melter observes the slag, and determines whether any conditioning material such as limestone, dolomite or coal must be added. High FeO levels in the slag (estimated by a simple correlation with turndown carbon and manganese values) are undesirable, and these slags are either not

splashed or are treated with conditioners to make them more viscous.

2. The vessel is rocked back and forth to coat the charge and tap pads. The furnace is then uprighted.

3. The nitrogen blow is started when the oxygen lance is lowered in the furnace to a predetermined height. The positioning of the lance and the duration of the nitrogen blow are determined by the melter, and depend on the condition of the slag and the area of the furnace that has to be coated.

4. At the end of the nitrogen blow, the furnace is tilted and excess slag is dumped into the slag pot.

Not only has slag splashing been successful in increasing the furnace lining life, but it has also significantly reduced the gunning costs. One major company has reported a reduction in gunning cost by 66%, and an increase in furnace availability from 78% to 97%. No detrimental effects on the turndown performance (chemical and thermal accuracy) of the furnace have been found.

Slag splashing has reduced the frequency of furnace relines considerably. This creates a problem in maintaining hoods, which is normally done during long furnace reline periods. Now, planned outages are needed for hood maintenance without any work on the furnace lining.

4.5.3.4 Effects of Desulfurized vs. Non-Desulfurized Hot Metal

Generally, heats that are made for sulfur sensitive grades of steel (0.010% or lower) must be desulfurized. Grades that do not have stringent sulfur specifications (approximately 0.018%) may not require hot metal desulfurization. The high sulfur content of the hot metal (approximately 0.06%) has to be reduced to acceptable levels in the hot metal desulfurizer, the BOF and/or in the ladle. The hot metal desulfurizing station can reduce the sulfur levels in the hot metal to as low as 0.001% through injection of lime and magnesium.

The BOF is not a very good desulfurizer because of the oxidizing conditions in the furnace. Heats using non-desulfurized hot metal can be desulfurized to approximately 0.012% sulfur in the BOF. However, heats using desulfurized hot metal may pick up sulfur if the slag in the desulfurizer was not skimmed properly or if a high sulfur scrap was used.

Some shops desulfurize virtually 100% of the hot metal to enable the blast furnace to run with a lower flux charge. This practice also reduces the heat burden on the blast furnace and saves coke. The net effect is a reduction in hot metal costs that is larger than the expense of desulfurizing.

4.6 Process Control Strategies

4.6.1 Introduction

Process control is an important part of the oxygen steelmaking operation as the heat production times are affected by it. Several steelmaking process control strategies are available today, and steel plants use strategies depending on their facilities and needs. Process control schemes can be broadly divided into two categories: static and dynamic. Static models determine the amount of oxygen to be blown and the charge to the furnace, given the initial and final information about the heat, but yield no information about the process variables during the oxygen blow. Static models are basically like shooting for a hole-in-one on the golf course: there is no further control once the ball leaves the face of the club. Dynamic models, on the other hand, make adjustments during the oxygen blow based on certain in-blow measurements. The dynamic system is like a guided missile in a military situation; there is provision for in-flight correction for better accuracy to hit the target.

Lance height control is an important factor in any control scheme as it influences the path of the process reactions. Detailed descriptions of static and dynamic models currently in use by the steel industry, along with a brief discussion of lance height control, follow.

4.6.2 Static Models

A static charge model is normally a computer program in an on-site steelmaking process computer. Plants have static models that depend on the type of operation and product mix. The static charge model uses initial and final information about the heat (e.g. the amount of hot metal and scrap, the aim carbon and temperature) to calculate the amount of charge and the amount of oxygen required. The relevant information is collected in the process computer, and a static charge model calculation is performed at the beginning of the heat. The output from the model determines the amount of oxygen to be blown and the amount of fluxes to be added to attain the desired (aim) carbon and temperature for that heat.

Performance of the charge model depends on a number of factors: the accuracy of the model; the accuracy of the inputs (aims, weights, chemistry of charge ingredients) to the process computer; consistency in the quality of materials used and of practices; and conformance to the static charge model.

The objective of the static charge model is to have improved control of carbon, temperature, sulfur and phosphorus. An accurate control over these variables using the model will result in fewer reblows and steel bath cooling, which are time consuming and expensive. Improved control will also result in better iron yield, greater productivity, better refractory lining life and an overall lower cost of production. The OBM (Q-BOP) static charge model is essentially the same as the BOF model, except for slight modifications arising out of differences in slag chemistry and natural gas injection. Most of the static charge models have an OK-to-tap performance of approximately 50%-80%.

There is one BOF shop in Europe that achieves over 90% OK-to-tap performance. It is a two-furnace LBE type shop with inert gas bottom stirring. They use only one furnace at a time while the other is being relined. Their success is attributable to discipline, attention to detail, pushing the one furnace very hard which results in consistent heat losses, almost zero furnace maintenance with resultant short lining lives, and well-developed computer models. Their resulting tap-to-tap times are very short and production rates are very high. There is no dynamic control system nor sensor sub-lance.

4.6.2.1 Fundamentals of the Static Charge Model

The charge model consists of a comprehensive heat and mass balance involving different chemical species that participate in the steelmaking reactions. The hot metal poured into the steelmaking vessel typically contains 4.5% C, 0.5% Si, 0.5% Mn and 0.06% P, apart from other impurities. These elements, along with Fe, get oxidized during the blow to form gases and slag, and these reactions generate the heat required to run the steelmaking process.

The static model also contains information on the heat of oxidation of different elements, such as carbon and iron, as a function of the oxygen blow. This information, along with a comprehensive mass balance involving different chemical species, determines the amount of heat generated from the steelmaking reactions as a function of total oxygen blown.

The primary coolants used in oxygen steelmaking processes are scrap, iron ore (sintered or lump), DRI and limestone (or dolomitic stone). The heat generated from the oxidizing reactions must be balanced with the use of these coolants to achieve the temperature aim of the heat. The model determines the amount of hot metal and scrap to be charged in the furnace, depending on the size and the type of heat to be made.

The rate of decarburization as a function of oxygen (available from the lance as well as from the ore) must be known to determine the duration of the blow, so that the heat turns down at the desired carbon level. The model also computes slag basicities (normally in the range of two to five) for effective dephosphorization and desulfurization. The amount of silica in the slag is calculated from hot metal silicon, assuming that it gets oxidized entirely to silica. The amount of lime (CaO) to be added is calculated from the desired basicity and the amount of silica formed from oxidation.

The charge model also determines the amount of dolomitic lime (CaO-MgO) to be charged into the furnace from an estimate of the overall slag chemistry. It is important to saturate the steelmaking slag with MgO to protect the furnace refractory lining.

4.6.2.2 Operation of the Static Charge Model

Many different kinds of charge models exist. In one typical model, charge calculations are made in three separate steps at the beginning of a heat. In the first step, called the hot metal calculation, the steelmaking operator makes a charge calculation of the amount of hot metal and scrap needed for the heat, given the product size, and the aim carbon and temperature of the heat. The hot metal analysis may or may not be available at this time. If not, an estimate based on the previous hot metal analysis is used.

Having fixed the hot metal and scrap weight, the operator runs the second part of the calculation, which is called the product calculation. For this, the hot metal analysis is required. The amount of fluxes to be added and the amount of oxygen to be blown are determined. This part of the calculation takes place before the start of the oxygen blow.

The third part, the oxygen trim calculation, takes place during the early stages of the oxygen blow and after the fluxes have been added to the furnace. In this step, the charge model utilizes the actual amounts of fluxes added to the furnace (which can be different from the calculated amounts), and provides the new trimmed value of total oxygen. This step is required to correct for any variations in the process due to weighing inaccuracies in the amount of flux charge.

4.6.3 Statistical and Neural Network Models

Over the last few years, a significant amount of attention has been given to statistical methods designed to improve the performance of the static charge model. The turn-down performance of the static model does not only depend on the accuracy of the mass and thermal balance of the steelmaking system, but also on the inherent variability of the process which arises out of a number of factors that cannot be accurately quantified. Steelmaking melters use the concept of bias adjustments to correct for unexplained inconsistencies in temperature or carbon performance. For example, if the melter finds a series of heats that were hot for unexplained reasons, he makes a temperature bias correction for the following heat, hoping that it will turn down close to the aim temperature. This bias correction, in effect, alters the aim temperature of the following heat for the purpose of charge calculation. The problem here is that different operators may react to a given situation differently and therefore the manual adjustments lack consistency. In addition, many operators are tempted to tamper (make adjustments too frequently), thereby introducing another variable.

Statistical models, in effect, use the same concept of bias to make adjustments in the charge calculation for a heat but do so with better consistency. The model tracks the turn-down performance of heats, and by statistical means, determines the amount of unexplained deviations in the performance of carbon, temperature etc. The model, then instructs the charge model to make appropriate modifications in the inputs for charge calculation for the following heat, so that these unexplained deviations may be countered. Studies indicate that the statistical models have been successful in fine tuning the static models to some extent.

Neural network algorithms are also being developed and used to improve the turndown performance. In this technique, the dependent variables (such as turndown carbon and temperature) are linked to the independent variables of a heat (such as the amount of hot metal, scrap and ore) through a network scheme. The network may consist of a number of layers of nodes, connected to each other in a linear fashion. As the input signals enter the network, they are multiplied by certain weights at the nodes and the products are summed as these signals are transferred from one layer to the next. The weights are determined by training the neural network model based on available historical data. The outputs from the model are calculated by adding the product of the weights and the signals in the last layer of the model.

Researchers have obtained reasonable success in estimating process variables using neural network modeling. However, the success of this technique depends on the quality and quantity of input signals that are fed into the model. In simple terms, the old computer proverb is still valid: "garbage in, garbage out."

4.6.4 Dynamic Control Schemes

Several steelmaking shops are currently using dynamic control schemes in combination with static models to improve their turndown performance. Dynamic control models use in-blow measurements of variables such as carbon and temperature, which can be used in conjunction with a static charge model, to fine tune the oxygen blow. A few dynamic control schemes are discussed in the following sections.

4.6.4.1 Gas Monitoring Schemes

This scheme is based on a continuous carbon balance of the entire BOF system. A gas monitoring system, which normally consists of a mass spectrometer, continuously analyzes dust free offgas samples for CO and CO_2 and determines the amount of carbon that has been oxidized at any given stage of the blow. Given the amount of carbon present in the system at the beginning of the heat (calculated from hot metal and scrap), one can calculate the level of carbon in the metal dynamically during the oxygen blow.

Although this method has a sound theoretical basis, it has not been very successful at the commercial level. One problem is that it is not possible to accurately determine the initial amount of carbon present in the hot metal and scrap. Even a small amount of error in determining this initial carbon content could result in serious process control problems. Another problem is the difficulty in continuous and accurate determination of the amount of CO and CO_2 exiting the system as offgas. Often the volume of filtering apparatus and tubing and the time to analyze a sample causes a delay of more than 60 seconds before the result is displayed or available for control.

4.6.4.2 Optical and Laser Based Sensors

A few shops in North America have started using light sensors to dynamically estimate carbon levels in the low carbon heats. The light meter system continuously measures the intensity of light emitted from the mouth of the steelmaking vessel during the oxygen blow. The system then correlates characteristics of the light intensity curve with the carbon content towards the end of the oxygen blow when the estimated carbon is around 0.06%. This system has been quite successful in dynamically estimating carbon levels in low carbon heats (0.06% or less), and adjusting the oxygen blow for improving the turndown carbon accuracy. However, this system cannot be used for heats with aim carbon greater than 0.06%.

A few oxygen steelmaking shops have tried using laser based sensors to dynamically estimate the bath temperature. Some laboratory researchers have attempted to devise a probe that would take in-blow metal samples and provide instant chemical analysis. However, most of these techniques are still in the developmental stage and are not yet commercially available.

4.6.4.3 Sensor or Sub-Lances

Sensor or sub-lances are effective tools for controlling both the carbon and temperature of the metal. In this technique a water-cooled lance containing expendable carbon and temperature sensors is lowered into the bath about two or three minutes before the end of the oxygen blow. The sensor determines the temperature of the metal bath at that time of blow and the carbon content of the steel by the liquidus arrest temperature method. The carbon and temperature readings thus taken are fed into the process computer which determines the additional amount of oxygen to be blown and the amount of coolants to be added.

The sensor lance has proven to be an extremely effective process control tool, with an OK-to-tap performance of 90% or more. The drawbacks of using sensor lances are substantial capital costs and engineering and maintenance problems. There are many oxygen steelmaking shops in Japan, and a few in North America, that are currently using the sensor lance technique to improve their turndown performance.

4.6.4.4 Drop-in Thermocouples for Quick-Tap

Drop-in thermocouples or bomb thermocouples are effective tools for measuring the temperature of the metal bath without turning down the steelmaking vessel. The thermocouple is contained in a heavy cylindrical casing, which is dropped into an upright vessel from the top of the furnace. The thermocouple probe has a sheathed wire attached to it which can convey a reliable emf reading before it burns up. The generated emf is conducted to a converter card and a computer, which converts the analog emf signal to a temperature value. The temperatures recorded by drop-in thermocouples have been found to be in good agreement with those measured using immersion thermocouples.

Drop-in thermocouples can be used to quick-tap heats, i.e. tap heats without turning the furnace down for chemical analysis. Certain grades of steel do not have stringent requirements for phosphorus or sulfur. In such a case, all that the steelmaking operator has to worry about at the end of the blow is temperature and carbon. If the operator feels confident that the carbon level in the bath is below the upper limit as specified by the grade (by either looking at the flame at the end of the oxygen blow, or by reading light meter carbon estimates, or by any other means), then he can use a drop-in thermocouple to determine the temperature of the metal bath without turning the vessel down. If the temperature recorded by the drop-in thermocouple is close enough to the aim temperature, then the heat can be quick-tapped.

Quick-tapping is becoming increasingly popular in North America with the advent of better devices for measuring in-blow carbon and bath temperatures. Quick-tapping has the obvious advantage of saving the production time that is otherwise consumed in turndowns. However, there is an obvious risk that a quick-tapped heat may be later found to have one or more of the elements present above their specification limits, in which case the heat may have to be regraded or scrapped. Another disadvantage is that the sensors cost between $0.05 to $0.10 per ton of steel.

4.6.4.5 Sonic Analysis

Several studies have been carried out in the past to correlate the audio emissions from the steelmaking furnace to the decarburization reaction and slag/foam formation. Investigators found an increase in the sound intensity at the beginning of the blow due to the establishment of the decarburization reaction. The sound intensity reportedly decreased with the onset of slag and foam formation in the slag. The sonic device can be used to control carbon and slag formation, which in turn can be used to control dephosphorization. These units can be affected by extraneous noise. Control is done by altering lance height or blow rate.

4.6.5 Lance Height Control

Lance height is defined as the vertical distance between the slag-metal interface in the furnace and the tip of the oxygen lance. The consequences of poor lance height control were discussed earlier in Section 4.2.2 on Oxygen Blow. It is an extremely important parameter as the height from which oxygen is blown affects the overall fluid flow of metal and slag during the blow. It is important to be able to measure it and to keep it consistent to achieve good process control of the furnace.

Measurement of lance height was also discussed earlier. However, the best practice in measuring lance height is to do it often. If the furnace bottom rapidly changes, due to either excess wear or to rapid bottom buildup, the lance height should be measured frequently (once per shift). If the furnace lining shape is stable, then less frequent determinations (once per day) will suffice. Lance height has a significant effect on slag formation, furnace and lance skull formation and decarburization rate. Generally, the safest, reasonably accurate method is calculating the bath height from refractory wear laser data.

4.7 Environmental Issues

4.7.1 Basic Concerns

Environmental or pollution control issues are becoming increasingly difficult and costly. State jurisdictions are setting increasingly stringent standards for new site permits or for significant modification to existing processes. Nearly every process has become a source of emissions. Usually, these must be controlled to a legal standard which is measured in pounds or tons per unit time. Sometimes, tradeoffs are permitted as long as a net major improvement or emissions reduction is achieved.

There are at least five major characteristic sources of environmental pollution. These include airborne emissions, water borne emissions, solid waste, shop work environment (usually burning and welding) and safety, and noise. Air emissions are the major issues in a BOF shop. Water borne pollutants, generated by the scrubber system, are clarified by settling and then the clean water is recirculated. Solid waste is generated by oxide bearing materials collected from the scrubber, electrostatic precipitators or baghouse. While much solid matter is recycled, the rest is put into long range storage for byproduct use (i.e., aggregate) and future recycle (for example, Waste Oxide Briquettes). Noise generally is not a major concern in a BOF (i.e., compared to an arc furnace).

Shop environmental and safety is always a major concern for personnel protection and employee morale. This discussion will concentrate principally on issues of air pollution.

4.7.2 Sources of Air Pollution

There are two broad areas: undesirable gases such as CO, fluorides or zinc vapors; and particulates such as oxide dusts. A brief summary of the various sources is discussed below.

4.7.2.1 Hot Metal Reladling

Pouring of hot metal from the torpedo car into the transfer ladle results in plumes of fine iron oxides and carbon flakes. This mixture, called kish is the major source of dust and dirt inside the shop. The graphite particles are generated because the liquid hot metal is saturated with carbon and when the temperature drops during pouring into the ladle, the carbon precipitates out as tiny graphite flakes. The usual method of control is to pour inside an enclosure (hood) and to collect the fume in a baghouse.

4.7.2.2 Desulfurization and Skimming of Hot Metal

Pneumatic injection of the liquid iron with nitrogen and magnesium with a lance stirs the metal and generates a fume similar to that of reladling. Often, the reladling hood is designed to accommodate the desulfurizing lance operation. During slag skimming, the splash of slag and metal is collected in a vessel (usually a slag pot) which also generates fine oxide fume. This is frequently collected in a separate hood over the skimming collection pot.

4.7.2.3 Charging the BOF

There is some fume generated when the scrap hits the bottom of the furnace. However, the major emission is generated while pouring hot metal into the furnace. Here, very dense oxide clouds, kish, and heavy flames rise quickly. Some shops have suction hoods on the charging side above the furnace that collect fume and divert the heat away from the crane. However, many shops are not so equipped and must rely on slow-pouring to limit the fume emission from the roof monitors to comply with regulations. Pouring too fast results in heavy flame reactions, that have been known to anneal the crane cables, causing spillage accidents.

4.7.2.4 Blowing (Melting and Refining)

By far the largest mass of fume is generated during the main blow. Approximately 90% by weight of total sources is generated at this point. The fume consists of hot gases at temperatures, over 1650°C (3000°F) and very heavy concentrations of iron oxide particles. The particulates can contain heavy metal oxides such as chromium, zinc, lead, cadmium, copper and others, depending on the scrap mix. The gas composition is approximately 80%-95% CO and the rest is CO_2. In open combustion hoods, where air is induced just above the furnace into the cleaning system, the temperatures can be as high as 1925℃ (3500°F) because of further combustion of CO. The fume is mostly fine iron oxides with some other oxides and dusts from flux additions.

The collection systems for this fume are of two types. Open hoods draft enough air to completely burn the CO before it hits the filtering device. Closed or suppressed combustion hoods either eliminate or reduce the induction of air to very low levels (<15%). This reduces the required fan horsepower and filtering capacity. However, consistent sealing between the furnace and hood is required to prevent the generating and igniting of explosive mixtures. Generally, the capital and operating costs for closed systems are less than for open systems. Thus, the most recently built shops are characterized by closed hood systems.

Another controlling characteristic is the filtering device. Two types have been successful in BOFs: electrostatic precipitators and venturi scrubbers.

The electrostatic precipitator draws the gas and fume into chambers. These are used only with open combustion hoods. These contain many parallel plates or alternating plates and wires spaced closely together within a few inches. Alternate elements (usually wires) are charged to a high potential. As the fume nears or contacts the highly charged element it becomes charged and is attracted to the other element (of opposite charge). Periodically the charge is dropped and the elements are vibrated, which releases the fume into a hopper and transport system below. Precipitators require gas temperatures less than 370℃ (700°F) and some moisture (>6%) to be efficient. A common problem is plate warpage, due to excessive heating, which results in electrical short-circuits.

The venturi scrubber, used with either open or suppressed combustion hoods, induces the gas/fume through a violent spray which washes and separates the fume from the gas. The oxide laden water is then subjected to a series of separation processes, which are a combination of centrifugal, chemical and settling operations. The settled filter cake is dredged and dried for recycling. Most facilities now use venturi scrubbers.

4.7.2.5 Sampling and Testing

Turning the furnace down for testing and temperature sampling generates fine oxide fumes at a relatively low rate. Most of this fume is drafted into the hood above. However, many shops running on limited fan capacity close off suction in the hood on the testing furnace and start blowing the next heat on the other furnace. The sampling/testing fume rates for a bottom-blown process are much heavier than top-blown furnaces due to continued blowing through the tuyeres to keep them clear. Most bottom-blown facilities use a charge-side enclosure, or doghouse, to effectively collect these emissions into the main hood. In effect, the doghouse is a set of doors that extends the hood down to the charging floor.

4.7.2.6 Tapping

Some fume comes directly from the furnace but most comes from steel colliding with the bottom of the ladle or other liquid steel. The addition of alloys increases the smoke during tap. Some fume is collected in the main hood, but much escapes through the roof monitors. Some shops use a tap collection hood and route the fume to a baghouse.

4.7.2.7 Materials Handling

The handling of fluxes, alloys and treatment reagents can be a significant source of fume. Often the handling facilities are equipped with a collection system that leads to a baghouse. Materials used in small amounts per heat can be transported in bags or super sacks to eliminate fugitive dust during transfer. Sometimes materials quality control specifications are required to control the generation of dust during handling and transfers. Reducing burnt lime fines specification is an example.

4.7.2.8 Teeming

Ingot teeming generally does not require special collection systems unless a hazardous element, such as lead, is involved. Free machining bar stock and plates alloyed with lead are common examples. Here lead is added while teeming ingots. A mobile hood collects the lead laden fume while teeming each ingot. It is both a hazardous material for the workers (vapors and particulate) and a solid waste. Further, workers' lead contaminated protective clothing and equipment are collected in special containers for proper disposal.

4.7.2.9 Maintenance and Skull Burning

The BOF operations generate skulls on equipment such as the furnace mouth, ladles, oxygen lances, metallic spills etc. These must be processed or removed by oxy-propane burning or oxygen lancing. Generally, this is a small emission source.

4.7.3 Relative Amounts of Fumes Generated

Most of the fume in the overall process is generated during the main blow. Table 4.11 summarizes the relative amounts of fume generated from various BOF sources. Overall, about 31lb (14kg) of fume are generated per ton of steel. Over 90% is from the main oxygen blow and reblows. The OBM (Q-BOP) and other major bottom blowing processes generate somewhat lesser amounts of fume with larger particle sizes which are easier to collect than the case of 100% top-blown. Desulfurizing and tapping each generate about one pound per ton (0.45kg/t).

Table 4.11 Summary of major emission sources of a BOF shop

Emission source approx.	Amt. /lb · t^{-1}(1lb=0.45kg, steel)
BOF hot metal transfer, source	0.190
Building monitor	0.056
Hot metal desulfurization, uncontrolled	1.090
Controlled by baghouse	0.009

Continued Table 4.11

Emission source approx.	Amt. /lb · t^{-1}(1lb=0.45kg, steel)
BOF charging, at source	0.600
Building monitor (roof top)	0.142
Controlled by baghouse	0.001
BOF refining, uncontrolled	28.050
Controlled with open hood, scrubber	0.130
Controlled with closed hood, scrubber	0.090
OBM (Q-BOP) refining, closed hood scrubber	0.056
BOF tapping, at source	0.920
Building monitor	0.290
Controlled by baghouse	0.003
Total at source or uncontrolled, BOF	30.850
Total controlled (scrubber, baghouse)	0.158
Total controlled (scrubber, monitors)	0.497

Table 4.11 also shows how much is removed by a reasonably good filtering system. A well designed system will reduce the fugitive emissions in a shop to less than 0.2lb/t(0.09kg/t), or by 99.4%. There is at least one consultant who has developed an effective modeling technique to test and optimize emission collection designs. A test is done in a water tank using small jets of salt water as pollution sources. The salt water is much denser than fresh water, and when put into a fresh water tank and modeled upside down, it generates a plume startlingly like the real thing. This technique has been very successful for both conformance to emission standards and for improving shop work environment. Most jurisdictions look favorably on such modeling to check the viability of candidate collection system designs.

4.7.4 Other Pollution Sources

4.7.4.1 BOF Slag

BOF slag is generated at a rate of approximately 240 pounds per ton (108.9kg/t) of steel. Currently about half of this quantity gets recycled within the plant to the sinter plant or is used directly in the blast furnace. Such in-plant slag recycling has been declining because of higher steel quality demands, e.g., lower phosphorus. Other uses such a aggregate and agricultural purposes are being explored.

4.7.4.2 BOF Dust and Sludge

BOF dust and sludge is generated at a rate of approximately 36 pounds per ton (16.3kg/t) of steel. Small amounts are sold to other industries, such as the cement industry, while small amounts are recycled to the sinter plant for metal recovery. There is a growing interest in using this material in the BOF as an iron source and coolant, in the form of WOBs or waste oxide briquettes.

The major driving forces for recycling come from government mandated pressures to reuse by-products and not to land fill them. In many cases there is an economic opportunity to use a low cost iron or coolant source. Problems to be overcome are buildup of phosphorus during recycling, zinc content, moisture, and the use of safe and suitable binders to improve handling characteristics.

4. 7. 5 Summary

Studying emission sources is a way to understand the process from another point of view. It is of increasing social and legal importance. Different jurisdictions do not always agree on acceptable levels, but they are all headed toward tougher environmental standards in the future. Accordingly, pollution control measures and evolving technologies will increasingly influence the design, capital, work environment, and operating costs of shops. Considerable thought by the steel industry is now being given to controlling emissions through engineering.

Exercises

4-1 Please fit the blanks to show the different operational steps.

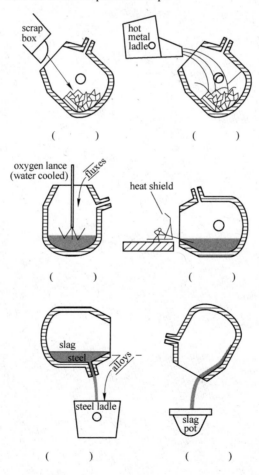

4-2 How many types of oxygen steelmaking processes, and what are they?

4-3 Please show the normal coolants for the oxygen steelmaking processes.

4-4 How reactions are there in BOF steelmaking process, please show them.

4-5 What is the significance of calculation for mass and energy balances of steelmaking?

4-6 Please discuss the process control strategies for steelmaking process.

4-7 What are the sources of air pollution for steelmaking process.

4-8 Please expound the significance of purification and recovery for converter emissions.

5 Electric Furnace Steelmaking

Over the past 30 years the use of the electric arc furnace (EAF) for the production of steel has grown considerably. There have been many reasons for this but primarily they all relate back to product cost and advances in technology. The capital cost per ton of annual installed capacity generally runs in the range of $140-200/t for an EAF based operation. For a similar blast furnace-BOF based operation the cost is approximately $1000 per annual ton of installed capacity. As a result EAF based operations have gradually moved into production areas that were traditionally made through the integrated route. The first of these areas was long products-reinforcing bar and merchant bar. This was followed by advances into heavy structural and plate products and most recently into the flat products area with the advancement of thin slab casting. At the current time, approximately 40% of the steel in North America is made via the EAF route. As the EAF producers attempt to further displace the integrated mills, several issues come into play such as residual levels in the steel (essentially elements contained in the steel that are not removed during melting or refining) and dissolved gases in the steel (nitrogen, hydrogen, oxygen).

Both of these have a great effect on the quality of the steel and must be controlled carefully if EAF steelmakers are to successfully enter into the production of higher quality steels.

There have been many advances in EAF technology that have allowed the EAF to compete more successfully with the integrated mills. Most of these have dealt with increases in productivity leading to lower cost steel production. These are described in the detailed process sections.

5.1 Electric Furnace Technology

The electric arc furnace has evolved considerably over the past 30 years. Gone are the days when electric power was the only source of energy for scrap melting. Previously tap-to-tap times in the range of 3-8 hours were common. With advances in technology it is now possible to make heats in under one hour with electrical energy consumptions in the range of 380-400 kW·h/t. The electric furnace has evolved into a fast and low cost melter of scrap where the major criterion is higher productivity in order to reduce fixed costs. Innovations which helped to achieve the higher production rates include oxy-fuel burners, oxygen lancing, carbon/lime injection, foamy slag practices, post-combustion in the EAF freeboard, EAF bath stirring, modified electrical supply (series reactors etc.), current conducting electrode arms, DC furnace technology and other innovative process technologies (scrap preheat, continuous charging etc.).

5.1.1 Oxygen Use in the EAF

Much of the productivity gain achieved over the past 10-15 years was related to oxygen use in the furnace. Exothermic reactions were used to replace a substantial portion of the energy input in the EAF. Whereas oxygen utilization of $9Nm^3/t$ (300scf/t) was considered ordinary just 10 years ago, some operations now use as much as $40Nm^3/t$ (1300scf/t) for lancing operations. With post-combustion, rates as high as $70Nm^3/t$ (2500scf/t) have been implemented. It is now common for 30%-40% of the total energy input to the EAF to come from oxy-fuel burners and oxygen lancing. By the early 1980s, more than 80% of the EAFs in Japan employed oxy-fuel burners. In North America it was estimated that in 1990, only 24% of EAF operations were using such burners. Since that time, a large percentage of North American operations have looked to the use of increased oxygen levels in their furnaces in order to increase productivity and decrease electrical energy consumption. High levels of oxygen input are standard on most new EAF installations.

The IISI 1990 electric arc furnace report indicates that most advanced EAF operations utilize at least $22Nm^3/t$ (770scf/t) of oxygen. In addition oxygen supplies 20%-32% of the total power input in conventional furnace operations. This has grown with the use of alternative iron sources in the EAF, many of which contain elevated carbon contents (1%-3%). In some cases, electrical energy now accounts for less than 50% of the total power input for steelmaking.

One of the best examples of the progressive increase in oxygen use within the EAF is the meltshop operation at Badische Stahlwerke (BSW). Between 1978 and 1990, oxygen use was increased from $9Nm^3/t$ to almost $27Nm^3/t$. Productivity increased from 32t/h to 85t/h while power consumption decreased from $494kW \cdot h/t$ to $357kW \cdot h/t$. During this period, BSW developed their own manipulator for the automatic injection of oxygen and carbon. In 1993, BSW installed the ALARC post-combustion system developed by Air Liquide and increased oxygen consumption in the furnace to $41.5Nm^3/t$. The corresponding power consumption is $315kW \cdot h/t$ with a tap-to-tap time of 48 minutes. This operation has truly been one of the pioneers for increased chemical energy use in the EAF.

5.1.2 Oxy-Fuel Burner Application in the EAF

Oxy-fuel burners are now almost standard equipment on electric arc furnaces in many parts of the world. The first use of burners was for melting the scrap at the slag door where arc heating was fairly inefficient. As furnace power was increased, burners were installed to help melt at the cold spots common to UHP operation. This resulted in more uniform melting and decreased the melting time necessary to reach a flat bath. It was quickly realized that productivity increases could be achieved by installing more burner power. Typical productivity increases reported in the literature have been in the range of 5%-20%. In recent years oxy-fuel burners have been of greater interest due to the increase in the cost of electrodes and electricity. Thus natural gas potentially provides a cheaper source of energy for melting. Fig. 5.1 shows oxy-fuel burners in operation in an EAF.

Typically burners are located in either the slag door, sidewall or roof. Slag door burners are generally used for small to medium sized furnaces where a single burner can reach all of the cold spots. Door burners have the advantage that they can be removed when not in use. For larger furnaces, three or four sidewall mounted burners are more effective for cold spot penetration. However, these are vulnerable to attack from slag, especially if employing a foamy slag practice. In such cases the burners are sometimes mounted in the roof and are fired tangentially through the furnace cold spots.

Fig. 5.1 Oxy-fuel burners in EAF operations

Oxy-fuel burners aid in scrap melting by transferring heat to the scrap. This heat transfer takes place via three modes; either forced convection from the combustion products, radiation from the combustion products or conduction from carbon or metal oxidation and from scrap to other scrap.

Primarily heat transfer is via the first two modes except when the burners are operated with excess oxygen. Heat transfer by these two modes is highly dependent on the temperature difference between the scrap and the flame and on the surface area of the scrap exposed for heat transfer. As a result oxy-fuel burners are most efficient at the start of a melt-in period when the scrap is cold. As melting proceeds the efficiency will drop off as the scrap surface in contact with the flame decreases and due to the fact that the scrap temperature also increases. It is generally recommended that burners be discontinued after 50% of the meltdown period is completed so that reasonable efficiencies are achieved. An added complication is that once the scrap heats up, it is possible for iron to react with the water formed by combustion to produce iron oxide and hydrogen. This results in yield loss and the hydrogen must be combusted downstream in the offgas system. Usually the point at which burner use should be discontinued is marked by a rise in offgas temperature(indicating that more heat is being retained in the offgas). In some operations the temperature of the furnace side panels adjacent to the burner is used to track burner efficiency. Once the efficiency drops below a set point the burners are shut off.

Heat transfer by conduction occurs when excess oxygen reacts with material in the charge. This will result in lower yield if the material burned is iron or alloys and as a result is not generally recommended. However in the period immediately following charging when volatile and combustible materials in the scrap flash off, additional oxygen in the furnace is beneficial as it allows this material to be burned inside the furnace and thus results in heat transfer back to the scrap. This is also beneficial for the operation of the offgas system as it not required to remove this heat downstream.

Fig. 5.2 indicates that 0.133MW of burner rating should be supplied per ton of furnace capacity. Other references recommend a minimum of 30kW · h/t of burner power to eliminate cold spots in a UHP furnace and 55-90kW · h/t of burner power for low powered furnaces.

A burner/lance has been developed by Empco (Fig. 5.3), that has the capability of following the scrap level as it melts. The lance is similar to a BOF design (water-cooled) and is capable of

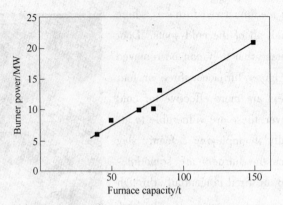

Fig. 5.2   Oxy-fuel burner rating versus furnace capacity

operation at various oxygen to natural gas ratios or with oxygen alone for decarburization. The lance is banana shaped and follows the scrap as its melts back. Thus a high heat transfer efficiency can be maintained for a longer period of time. Typical efficiencies reported for the Unilance are in the range of 65%-70%.

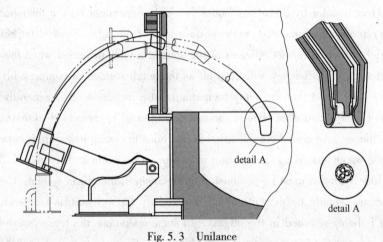

Fig. 5.3   Unilance

Heat transfer efficiencies reported in the literature vary greatly in the range of 50%-75%. Data published by L'Air Liquide shows efficiencies of 64%-80% based on energy savings. Krupp indicates an energy efficiency based on energy savings alone of 78%, and also gives an indication of why such high efficiencies are reported when looking at energy replacement alone. According to Krupp, an increase of one minute in meltdown time corresponds to an increase in power consumption of $2kW \cdot h/t$. Thus the decrease in tap-to-tap time must be taken into account when calculating burner efficiency. For typically reported energy efficiencies, once this is taken into account burner efficiencies lie in the range of 45%-65% in all cases. This is supported somewhat by theoretical calculations by Danieli that indicated that only 20%-30% of the energy from the burners went to the scrap; as much as 40%-60% of the heat went to the offgas and the remainder was lost to the water-cooled furnace panels.

Fig. 5.4 shows burner efficiency as a function of operating time based on actual furnace offgas measurements. This shows that burner efficiency drops off rapidly after 40%-50% of the melting time. By 60% into the melting time burner efficiency has dropped off to below 30%. It is apparent that a cumulative efficiency of 50%-60% is achieved over the first half of the meltdown period and drops off rapidly afterwards. As a result, typical operating practice for a three bucket charge is to run the burners for 2/3 of the first meltdown, 1/2 of the second meltdown and 1/3 of the third meltdown. For operations with only one backcharge, burners are typically run for 50% of each meltdown phase.

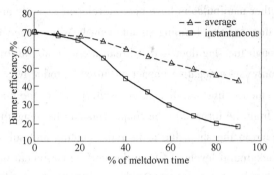

Fig. 5.4 Oxy-fuel burner efficiency versus time into meltdown

The amount of power input to the furnace has a small effect on the increase in heatload and offgas volume. The major factor is the rate at which the power is put into the furnace. Thus for a low powered furnace (high tap-to-tap time), the total burner input to the EAF may be in the range of 80-100kW · h/t but the net effect on the offgas evacuation requirements may not change much as the burners are run for a longer period of time. Likewise a high powered furnace, due to the short tap-to-tap time, the burner input rate may be quite high even though the burners supply only 30kW · h/t to the furnace.

Bender estimates that offgas flowrates can increase by a factor of 1.5 and offgas heatload by as high as a factor of 2.5 for operation with burners and high oxygen lance rates.

5.1.3 Application of Oxygen Lancing in the EAF

Over the past ten years oxygen lancing has become an integral part of EAF melting operations. It has been recognized in the past that productivity improvements in the open hearth furnace and in the BOF were possible through the use of oxygen to supply fuel for exothermic reactions. Whereas previously oxygen was used primarily only for decarburization in the EAF at levels of 3-8Nm^3/t (96-250scf/t), in modern operations anywhere from 10%-30% of the total energy input is supplied via exothermic bath reactions. For a typical UHP furnace oxygen consumption is in the range of 18-27Nm^3/t (640-960scf/t). Oxygen utilization in the EAF is much higher in Japan and in Europe where electricity costs are higher. Oxygen injection can provide a substantial power input—a lance rate of 30Nm^3/min (1050scfm) is equivalent to a power input of 11MW based on the theoretical reaction heat from combustion of C to CO.

Oxygen lances can be of two forms. Water-cooled lances are generally used for decarburization though in some cases they are now using for scrap cutting as well. The conventional water-cooled lance was mounted on a platform and penetrated into the side of the furnace through a panel. Water-cooled lances do not actually penetrate the bath though they sometimes penetrate into the slag layer. Consumable lances are designed to penetrate into the bath or the slag layer. They consist of consumable pipe which is adjusted as it burns away to give sufficient working length. The first consumable lances were operated manually through the slag door. Badische Stahl Engineering developed a robotic manipulator to automate the process. This manipulator is used to control two lances automatically. Various other manipulators have been developed recently and now have the capability to inject carbon and lime for slag foaming simultaneously with oxygen lancing. One major disadvantage of lancing through the slag door is that it can increase air infiltration into the furnace by 100%-200%. This not only has a negative impact on furnace productivity but also increases offgas system evacuation requirements substantially. As a result, not all of the fume is captured and a significant amount escapes from the furnace to the shop. This can be a significant problem if substantial quantities of CO escape to the shop due to its rapid cooling and subsequent incomplete combustion to CO_2. Thus background levels of CO in the work environment may become an issue. To reduce the amount of air infiltration to the EAF, some operations insert the lance through the furnace sidewall as shown in Fig. 5.5.

Energy savings due to oxygen lancing arise from both exothermic reactions (oxidation of carbon and iron) and due to stirring of the bath with leads to temperature and composition homogeneity of the bath. The product of scrap cutting is liquid iron and iron oxide. Thus most of the heat is retained in the bath. The theoretical energy input for oxygen reactions in the bath is as follows:

$$Fe + 1/2 O_2 = FeO, \text{ heat input} = 6.0 kW \cdot h/Nm^3 O_2 \quad (5.1)$$

$$C + 1/2 O_2 = CO, \text{ heat input} = 2.8 kW \cdot h/Nm^3 O_2 \quad (5.2)$$

Thus it is apparent that much more energy is available if iron is burned to produce FeO. Naturally though, this will impact negatively on productivity. Studies have shown that the optimum use of oxygen for conventional lancing operations is in the range of 30-40 Nm^3/t (1000-1250 scf/t). Above this level yield losses are excessive and it is no longer economical to add oxygen. Typical operating results have given energy replacement values for oxygen in the range of 2-4 $kW \cdot h/Nm^3 O_2$ (0.056-0.125 $kW \cdot h$/scf O_2), with an average of 3.5 $kW \cdot h/Nm^3 O_2$ (0.1 $kW \cdot h$/scf O_2). These values show that it is likely that both carbon and iron are reacting. In addition, some studies have shown that the oxygen yield (i.e. the amount reacting with carbon) is in the range of 70%-80%. This would support the theory that both carbon and iron are reacting. During scrap cutting operations, the oxygen reacts primarily with the iron. Later when a molten pool has formed the FeO is reduced out of the slag by carbon in the bath. Thus the net effect is to produce CO gas from the oxygen that is lanced.

Based on the information cited in the preceding section, it can be expected that for every Nm^3 of oxygen lanced, 0.75 Nm^3 will react with carbon to produce 1.5 Nm^3 of CO (based on the average energy replacement value of 3.5 $kW \cdot h/Nm^3 O_2$). If in addition the stirring effect of the lancing brings bath carbon or injected carbon into contact with FeO in the slag, an even greater quantity of

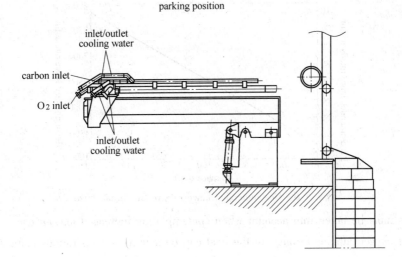

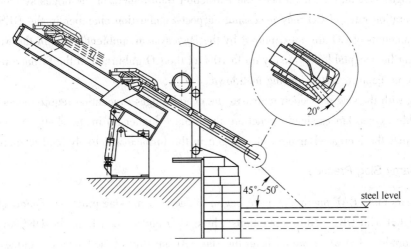

Fig. 5.5 Operation of oxygen lance through the furnace sidewall (Courtesy of Danieli.)

CO may result. That this occurs is supported by data that indicates a decarburization efficiency of greater than 100%. Thus during the decarburization period up to 2.5Nm^3 of CO may result for every Nm^3 of oxygen injected. Typical oxygen rates are in the range of 30-100Nm^3/min (1000-3500scfm) and are usually limited by the ability of the fourth hole system to evacuate the furnace fume. Recommended lance rates for various furnace sizes are shown in Fig. 5.6 and indicate a rate of approximately 0.78-0.85Nm^3/t (25-30 scfm/t) of furnace capacity. In some newer processes where feed materials are very high in carbon content, oxygen lance rates equivalent to 0.1% decarburization per minute are required. In such cases, the lance rates may be as high as 280Nm^3/min (10,000scfm) which is similar to BOF lance rates.

The major drawback to high oxygen lance rates is the effect on fume system control and the production of NO$_x$. Offgas volumes are greatly increased and the amount of CO generated is much

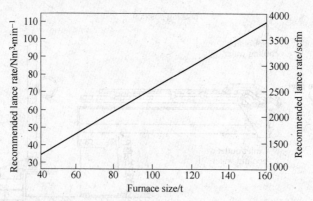

Fig. 5.6 Recommended lance rate versus furnace size

greater. This must be taken into account when contemplating increased oxygen use.

The use of oxygen lancing throughout the heat can be achieved in operations using a hot heel in the furnace. Oxygen is lanced at a lower rate throughout the heat to foam the slag. This gives better shielding of the arc leading to better electrical efficiency. It also gives lower peak flowrates of CO to the offgas system, thus it reduces the extraction requirement of the offgas system.

High generation rates of CO may necessitate a post-combustion chamber in the DES system. If substantial amounts of CO are not captured by the DES system, ambient levels in the work environment may not be acceptable. Typically up to 10% of the CO unburned in the furnace reports to the secondary fume capture system during meltdown.

Operating with the slag door open increases the overall offgas evacuation requirements substantially. If possible oxygen lances should penetrate the furnace higher up in the shell. Another factor to consider is that the increased amount of nitrogen in the furnace will likely lead to increased NO_x.

5.1.4 Foamy Slag Practice

In recent years more EAF operations have begun to use a foamy slag practice. Foamy slag was initially associated with DRI melting operations where FeO and carbon from the DRI would react in the bath to produce CO which would foam the slag. At the start of meltdown the radiation from the arc to the sidewalls is negligible because the electrodes are surrounded by the scrap. As melting proceeds the efficiency of heat transfer to the scrap and bath drops off and more heat is radiated from the arc to the sidewalls. By covering the arc in a layer of slag, the arc is shielded and the energy is transferred to the bath as shown in Fig. 5.7.

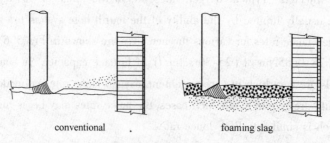

Fig. 5.7 Effect of slag foaming on arc radiation (Courtesy of Center for Materials Production)

Oxygen is injected with coal to foam up the slag by producing CO gas in the slag. In some cases only carbon is injected and the carbon reacts with FeO in the slag to produce CO gas. When foamed, the slag cover increases from 4 inch (10.16cm) thick to 12 inch (30.48cm). In some cases the slag is foamed to such an extent that it comes out of the electrode ports. Claims for the increase in efficiency range from an efficiency of 60%-90% with slag foaming compared to 40% without. This is shown in Fig. 5.8 and Fig. 5.9. It has been reported that at least 0.3% carbon should be removed using oxygen in order to achieve a good foamy slag practice. If a deep foamy slag is achieved it is possible to increase the arc voltage considerably. This allows a greater rate of power input. Slag foaming is usually carried out once a flat bath is achieved. However, with hot heel operations it is possible to start slag foaming much sooner.

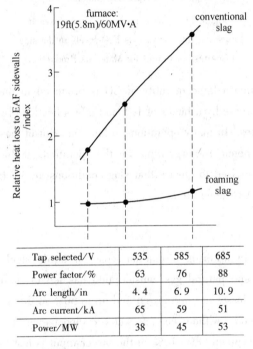

Tap selected/V	535	585	685
Power factor/%	63	76	88
Arc length/in	4.4	6.9	10.9
Arc current/kA	65	59	51
Power/MW	38	45	53

Fig. 5.8 Effect of slag foaming on heat loss to the furnace sidewalls (Courtesy of Center for Materials Production)

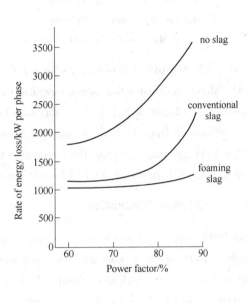

Fig. 5.9 Effect of slag cover on heat loss in the EAF (Courtesy of Center for Materials Production)

Some of the benefits attributed to foamy slag are decreased heat losses to the sidewalls, improved heat transfer from the arcs to the steel allows for higher rate of power input, reduced power and voltage fluctuations, reduced electrical and audible noise, increased arc length (up to 100%) without increasing heat loss and reduced electrode and refractory consumption.

Several factors have been identified that promote slag foaming. These are oxygen and carbon availability, increased slag viscosity, decreased surface tension, slag basicity > 2.5 and FeO in slag at 15%-20% to sustain the reaction. Fig. 5.10 and Fig. 5.11 show several of these effects graphically.

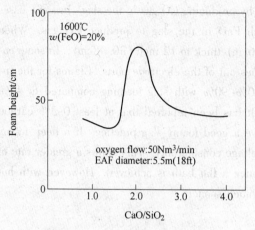
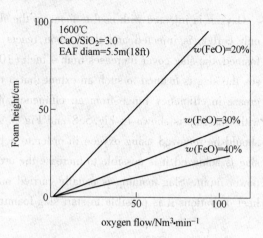

Fig. 5.10 Slag foam height versus slag basicity (Courtesy of Center for Materials Production)

Fig. 5.11 Slag foam height versus oxygen injection rate for various FeO levels in the slag (Courtesy of Center for Materials Production)

The only negative side of foamy slag practice is that a large quantity of CO is produced in the EAF. Bender estimates that offgas flowrates can increase by a factor of 1.5 and offgas heatload by as much as a factor of 2.5 for high slag foaming rates. In many operations, a large amount of carbon is removed from the bath in order to generate chemical energy input for the operation. If this CO is to be generated anyway, the operator is well advised to ensure that slag conditions are such that foaming will result in order to benefit from the CO generation.

5.1.5 CO Post-Combustion

The 1990s have seen steelmakers advance further in lowering production costs of liquid steel. Higher electrical input rates and increased oxygen and natural gas consumption has led to short tap-to-tap times and high throughputs. Thus, energy losses are minimized and up to 60% of the total power input ends up in the steel. This has not come without cost, as water-cooled panels and roofs are required to operate at higher heat fluxes. Typically 8%-10% of the power input is lost to the cooling water and offgas temperatures are extremely high, with losses of approximately 20% of the power input to the offgas. As EAF steelmakers attempt to lower their energy inputs further they have begun to consider the heat contained in the offgas. One way in which this can be recaptured is to use the offgas to preheat scrap. This results in recovery of the sensible heat but does not address the calorific heat which can represent as much as 50%-60% of the energy in the offgas.

5.1.5.1 Introduction

Generically, post-combustion refers to the burning of any partially combusted compounds. In EAF operations both CO and H_2 are present. CO gas is produced in large quantities in the EAF both from oxygen lancing and slag foaming activities and from the use of pig iron or DRI in the charge. Large amounts of CO and H_2 are generated at the start of meltdown as oil, grease and other combustible materials evolve from the surface of the scrap. If there is sufficient oxygen present, these compounds will burn to completion. In most cases there is insufficient oxygen for complete combustion and high levels of CO result. Tests conducted at Vallourec by Air Liquide showed that the offgas from the furnace could contain considerable amounts of non-combusted CO when there was insufficient oxygen present.

The heat of combustion of CO to CO_2 is three times greater than that of C to CO (for dissolved carbon in the bath). This represents a very large potential energy source for the EAF. Studies at Irsid (Usinor SA) have shown that the potential energy saving is significant and could be a much as 72kW·h/t. If the CO is burned in the freeboard it is possible to recover heat within the furnace. Some of the expected benefits and concerns regarding post-combustion are given in Table 5.1. As more oxygen is used to reduce electrical consumption, there will be greater need for improved oxygen utilization. In addition, environmental regulations may limit CO_2 emissions. As a result it will be necessary to obtain the maximum benefit of oxygen in the furnace. This can only be achieved if most CO is burned in the furnace.

Table 5.1 Benefits and concerns for post-combustion

Benefits	Concerns
Decreased heat load to the offgas system	Increased electrode consumption
Decreased CO emissions to the meltshop and baghouse	Increased heat load to the water-cooled panels and roof
Higher heat transfer due to higher radiation from combustion products	Decreased iron yield
Decreased water-cooled duct requirement	Economics of additional oxygen
Increased utilization of energy from oxygen and carbon	
Reduced electrical power consumption	
Decreased NO_x emissions from the EAF	
Increased productivity without increased offgas system requirements	

In order to maintain consistency of the results presented the following definitions are made for EAF trials:

$$\text{Post combustion ratio (PCR)} = \frac{CO_2}{CO+CO_2} \quad (5.3)$$

$$\text{Heat transfer efficiency (HTE)} = \frac{\text{reduction in kW·h to steel}}{\text{theoretical energy of PC for CO}} \quad (5.4)$$

5.1.5.2 Post-Combustion in Electric Arc Furnaces

Several trials have been run using post-combustion in the EAF. In some of the current processes, oxygen is injected into the furnace above the slag to post-combust CO. Some processes involve injection of oxygen into the slag to post-combust the CO before it enters the furnace freeboard. Most of these trials were inspired by an offgas analysis which showed large quantities of CO leaving the

EAF. Fig. 5.12 shows a typical post-combustion system where oxygen is added in the freeboard of the furnace. Thus combustion products directly contact the cold scrap. Most of the heat transfer in this case is radiative. Fig. 5.13 shows the approach where post-combustion is carried out low in the furnace or in the slag itself. Heat transfer is accomplished via the circulation of slag and metal droplets within the slag. Post-combustion oxygen is introduced at very low velocities into the slag. Heat transfer is predominantly convective for this mode of post-combustion. Some other systems have incorporated bottom blown oxygen (via tuyeres in the furnace hearth) along with injection of oxygen low

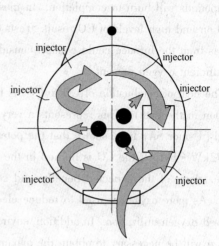

Fig. 5.12 Injection of postcombustion oxygen above the slag (Courtesy of Air Liquide)

in the furnace. In fact the first two processes to employ extensive post-combustion as part of the operation (K-ES, EOF) both took this approach.

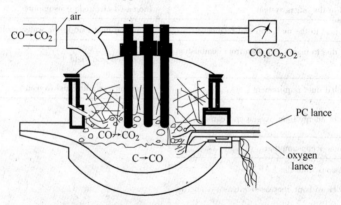

Fig. 5.13 Post-com-bustion in the slag layer
Maximimum degree of post combustion and heat transfer
PC in scrap melting and flat bath periods
Unit power savings > 4kW·h/Nm³ O_2
8%-10% power savings, 6%-8% productivity increase

Some of the results that have been obtained are summarized in Table 5.2 along with their theoretical limits. Note that in many cases the energy savings has been stated in kW·h/Nm³ of post-combustion oxygen added. The theoretical limit for post-combustion of CO at bath temperatures (1600℃) is 5.8kW·h/Nm³ of oxygen. PCRs have been calculated based on the offgas analysis presented in these references. HTE is rarely measured in EAF post-combustion trials and is generally backed out based on electrical energy replacement due to post-combustion. This does not necessarily represent the true post-combustion HTE because the oxygen can react with compounds other than CO.

5.1 Electric Furnace Technology

Table 5.2 Typical post-combustion results for scrap based operations

Location	Method	Net Efficiency/%	HTE/%	PCR/%	kW·h/Nm³ O_2
Theoretical		100			5.8
Acciaierie Venete SpA	K-ES	40-50		80	
Commonwealth Steel Co Ltd	Independent	30			
Ferriere Nord SpA	K-ES/DANARC	40	50+	72-75	3-4
Vallourec	Air Liquide	60		66-71	
Ovako Steel AB	AGA	61	77-87		4.3
Nucor	Praxair	80 (lance)	80	42-75	5.04
		60 (burners)			3.78
UES Aldwerke	Air Products/independent	64			3.7
Von Roll (Swiss Steel AG)	Air Liquide				2.53
Ferriera Valsabbia SpA	Air Liquide	Burners			1.2-1.6
		ALARC PC			3.0-3.5
Atlantic Steel	Praxair			60-70	3.7-5.13
Cascade Steel	Air Liquide				2.5-3.0
Badische Stahlwerke GmbH	Air Liquide	64		83	3.69
Cia Siderurgica Pains SA	Korf EOF	60		80-100	

5.1.5.3 Theoretical Analysis of Post-Combustion

Due to the scatter of reported results to date for EAF post-combustion it is necessary to discuss the theoretical analysis to put the results into perspective. In addition, the trials in converters and smelters can help to draw conclusions.

A Post-Combustion and Heat Transfer Mechanisms

Heat transfer to a mass of solid scrap is relatively efficient due to its large surface area and the inherent large difference in temperature. However, heat transfer to the bath is more difficult and two methods are proposed. Interaction of metal and slag droplets occurs in the Hismelt process while heat transfer is through the slag phase in deep slag (DIOS and AISI) processes. In bath smelting, post-combustion energy is consumed and is not transferred to the bath. Exact mechanisms are not known at this time.

a Post-Combustion in the Furnace Freeboard

All of the post-combustion results indicate that high efficiency can only be achieved by effectively transferring the heat from the post-combustion products to some other material. This has been verified in several BOF post-combustion trials where scrap additions during post-combustion resulted in higher post-combustion levels. If the combustion products do not transfer their heat, they will tend to dissociate. The best opportunity during scrap melting is to transfer the heat to cold scrap.

The adiabatic flame temperature for the combustion of CO with oxygen levels off at 2800℃. The adiabatic flame temperature does not increase proportionally to the amount of CO burned. Thus the amount of heat transferred from the flame does not increase proportionally. The adiabatic flame temperature for oxygen/natural gas combustion is in the range of 2600-2700℃. Natural gas combustion produces a greater volume per unit oxygen and therefore more contact with the scrap should occur. Trials with oxy-fuel and oxy-coal burners indicate that efficiencies of 75%-80% can be obtained at the start of meltdown, but that efficiency drops off quickly as the scrap heats and an average efficiency of 65% can be expected. For post-combustion, the efficiency can be expected to be similar or lower, depending on how long the process is carried out. Burner efficiency over a flat bath is 20%-30%, setting a lower limit and as a result, heat flux to the furnace panels will increase. Air Liquide has reported PCR = 70% and HTE = 50% for flat bath operations with post-combustion. This equates to a net efficiency of 35%.

Increased electrode consumption is a possibility. High bath stir rates may help to recover more of this energy. If the heat is not transferred from the post-combustion gases then a high PCR cannot be expected since some of the CO_2 will dissociate. During scrap melting oil and grease burn off the scrap. CO and H_2 concentration spikes occur when scrap falls into the bath. The hydrogen from these organics tend to form hydoxyl ions which enhance the rate at which CO can be postcombusted. The reaction between CO and O_2 is believed to be a branched chain reaction whose rate is greatly enhanced if a small quantity of H_2 is present. Thus higher PCRs in the EAF should be achieved during scrap meltdown. This benefit will not occur for post-combustion carried out in the slag though oxygen injected into the slag will continue to react in the furnace freeboard once it leaves the slag.

Nippon Steel found in their trials that for high offgas temperatures, radiation was the dominant heat transfer mechanism but the net amount of heat that was transferred was low. For temperatures lower than 1765℃, heat transfer by radiation and convection accounted for 30% of the heat transfer. The remainder was due to circulation of materials in the slag.

b Post-Combustion in the Slag Layer

Praxair has taken the approach of using subsonic or soft blowing of post-combustion oxygen through a multi-nozzle lance. The post-combustion lance is positioned close to the primary oxygen injection lance. The impact of the supersonic primary oxygen jet produces an emulsion of metal droplets in the slag. For post-combustion operations in an EAF, Praxair estimates that 0.25% of the steel bath needs to be emulsified in order to transfer all of the post-combustion heat. Praxair has found no evidence that post-combustion oxygen reacts with the metal droplets or with carbon in the slag. Post-combustion is carried out both during scrap melting and during a flat bath. During the start of scrap melting, post-combustion is difficult to achieve because oxygen lances (especially water-cooled lances) cannot be introduced into the furnace until sufficient scrap has meted back from the slag door. Information supplied by Praxair indicates that they do not attempt to burn all of the CO which is generated. Rather, they attempt to burn only a portion of the CO ranging from 10%-50% of that which is generated from primary oxygen lancing.

The injection of oxygen must be localized and well controlled to avoid oxygen reaction with the electrodes. Previous trials for post-combustion in EAF slags have used a piggyback lance over top of the decarburizing oxygen lance. Thus an attempt is made to burn the CO in the slag as it leaves the bath.

The slag layer must be thick in order to avoid the reaction of CO_2 with the bath to produce CO which would make a high PCR difficult to achieve. In converter and bath smelting operations, the slag layer is considerably thicker than in the EAF. Sumitomo proposed that a deep slag layer be used with post-combustion in the upper layer of the slag where the oxygen would not react with metal droplets. The temperature gradient in the slag layer was minimized with side blown oxygen into the slag. From Nippon Steel indications were that moderate bath stirring is necessary to accelerate reduction of iron ore and to enhance heat transfer from post-combustion. A thick slag layer was used in order to separate the metal bath from the oxygen and the post-combustion products. A thick slag layer also helped to increase post-combustion and decrease dust formation. It might prove economical to couple slag post-combustion in the EAF with addition of DRI/HBI following scrap melting operations although DRI addition typically results in much slag foaming.

Post-combustion could take place in the upper layer of the slag. Intense stirring would transfer heat back to the slag/bath interface where the DRI/HBI would be melted. This could lead to optimum energy recovery for post-combustion in the slag. Bath smelting results indicate that a net efficiency of about 50% can be achieved when CO is post-combusted in the slag.

During scrap melting operations, post-combustion above the slag low in the furnace freeboard should give the best results. During flat bath operations, it may be advantageous to burn the CO in or just above the slag. However, post-combustion in the slag will be a much more complicated process and will require a greater degree of control over the post-combustion oxygen dispersion pattern in the slag in order to prevent undesirable reactions.

B   Electrode Consumption

Previous studies have attributed electrode consumption to two mechanisms: sidewall consumption and tip erosion. Tip erosion is dependent on the electrical current being carried by the electrode. Sidewall consumption is dependent on the surface temperature of the electrode and can be controlled by providing water cooling of the electrode (sprays). During the initial stages of meltdown, the scrap helps to shield the electrodes and the hot surface will not react with oxygen present in the furnace. Once the scrap melts back from the electrodes, the hot carbon surface will react with any oxygen that is present. The source of the oxygen can be injected oxygen or oxygen that is present in air that is pulled into the EAF through the slag door.

If the electrodes are not shielded from the combustion products of post-combustion, carbon can react with CO_2 to form CO. The electrode will also react with any available oxygen. Both mechanisms will increase electrode consumption. At high offgas temperatures, the amount of electrode consumption can increase drastically.

C   Heatload to the EAF Shell

The usefulness of the heat generated by post-combustion will be highly dependent on the effective heat transfer to the steel scrap and the bath. The theoretical adiabatic flame temperature for CO combustion with oxygen (2800℃) is similar to that for combustion of natural gas with oxygen. It is well known that when oxy-fuel burners are fired at very high rates on dense scrap, the flame can blow back against the water-cooled panels and overheat them. If the heat from post-combustion is not rapidly transferred to the scrap, overheating of the panels will also occur. If the combustion products do not penetrate the scrap, then most of the heat transfer will be radiative and the controlling rate mechanism will be conductive heat transfer within the scrap. This will be very slow compared to the radiative heat transfer to the scrap surface and will result in local overheating.

D Iron Yield

Oxy-fuel burner use can lead to yield losses and increased electrode consumption as some combustion products react with iron to form FeO. Trials run by Leary and Philbrook on the preheating of scrap with oxy-fuel burners showed that above scrap temperatures of 760℃, 2%-3% yield loss occurred. This is supported by work that shows that the thermodynamic equilibrium between iron and CO_2 at temperatures greater that 1377℃ is 24% CO_2. Gas temperatures exceeding 1800℃ are possible in the EAF and the corresponding equilibrium at this temperature is 8.6% CO_2. Though the gas residence time in the EAF probably will not allow for equilibrium to be reached, some oxidation will occur. If additional carbon is not supplied a yield loss will occur. This is likely to be the case for post-combustion. An iron yield loss of 1% equates to a power input of 12kW · h/t. This can have a significant effect on the overall post-combustion heat balance resulting in a fictitiously high HTE for post-combustion.

As Fe is oxidized to FeO, a protective layer can form on the scrap. Once the FeO layer is formed, oxygen must diffuse through the layer in order to react with the iron underneath. This will help to protect the scrap from further oxidation if this layer does not peel off exposing the iron. At temperatures above 1300℃ the FeO will tend to melt and this protective layer will no longer exist.

E Limits on Potential Gains from Post-Combustion

Bender indicates that CO emissions typically range between 0.3 and 2.6kg/t. Air infiltration into the furnace will result in natural post-combustion. This is indicated in several EAF post-combustion studies where CO_2 levels prior to the trials are in the range of 15%-40% and PCR ranges from 30%-60%. With a postcombustion system, CO levels typically dropped to below 10%. The starting level of CO has a big effect on the efficiency obtained once post-combustion is installed. These results are presented in Table 5.3.

Table 5.3 PCR with and without injection of post-combustion oxygen

Location	PCR prior to post-combustion of O_2/%	PCR after post-combustion of O_2/%
Vallourec	33-50	66-71
BSW	58-61	83
Nucor Plymouth	30-50	42-75
UES Aldewerke	40-62.5	

Continued Table 5.3

Location	PCR prior to post-combustion of O_2/%	PCR after post-combustion of O_2/%
Ferrierre Nord		71-75
Venete		80
Co-Steel Sheerness	22-62.5	
AGA	44-66	77-87

CO_2 has a higher heat capacity than CO. As a result for the same offgas temperature, the CO_2 will remove more heat per Nm^3. Thus a portion of the post-combustion energy will be removed from the furnace in the CO_2. For an offgas temperature range of 1200-1400℃ the maximum HTE is limited to 93.6%-92%. If the slag is heated by 110℃ the heat contained is equivalent to 0.9-1.2kW · h/t (assumes 4%-5% slag). These factors should be taken into account.

Some benefits are generally unaccounted for when post-combustion is implemented. A decrease of one minute in tap-to-tap time can save 2-3kW · h/t. A more recent study indicates a savings of 1kW · h/t for highly efficient furnaces. This is dependent on the efficiency of the current operation. Decreased cycle times should be considered when backing out the true heat transfer efficiency.

F Environmental Benefits

Increased volumes of injected gas will tend to decrease the amount of furnace infiltrated air. A positive pressure operation can save from 9-18kW · h/t and some savings can be expected to result from a reduction in air infiltration depending on the operation of the offgas system. However, a reduction in NO_x due to reduced air infiltration has yet to be shown in EAF post-combustion operations. If we consider the amount of air infiltration at the start of meltdown we can see that the scrap will help to retard airflow into the furnace. Thus at the start of meltdown, perhaps the effective opening area is only 25% of the slag door area. By the end of meltdown this area is almost entirely open so now the effective opening might be 90% of the slag door area. Thus the amount of air infiltration into the furnace increases drastically over the course of the meltdown period. This will have a big affect on the effectiveness of auxilliary postcombustion oxygen added to the furnace.

Post-combustion helps the fume system capacity because the amount of heat that must be removed by the water-cooled duct is decreased. Several operations report decreased offgas temperatures when using post-combustion. CO emissions at the baghouse are a function of offgas system design (sizing of the combustion gap, evacuation rate). Burning of some of the CO in the EAF will help to ensure that less CO enters the canopy system which will help to reduce CO emissions. Lower CO emissions have been reported in several EAF post-combustion operations. If CO levels are reduced below 5% in the furnace, it may be difficult to complete CO combustion in the offgas system following addition of dilution air at the combustion gap. Thus controlling CO levels leaving the furnace to approximately 10% is a good practice. Alternatively a post-combustion chamber may be used within the offgas system. It has been shown that CO tends to react with NO at high temperatures to form nitrogen and carbon dioxide as follows:

$$2CO+2NO = 2CO_2+N_2 \tag{5.5}$$

Thus it is beneficial to have a certain amount of CO present in the furnace in order to reduce the NO emissions. This is another good reason why total post-combustion of the CO is not desirable.

Trials carried out at Dofasco on their K-OBM showed that when excessive slag foaming occurred, particulate emissions from the furnace increased by as much as 59%. Nippon Steel found that a thick slag layer helped to increase post-combustion and decrease dust formation. For post-combustion in the slag in the EAF, high dust generation rates should be expected unless a thick slag layer is used. Excessive stirring of the bath will also lead to increased dust generation if metal droplets react with the post-combustion oxygen.

G Need for a Post-Combustion Chamber

If an attempt is made to burn close to all of the CO in the furnace, it is likely that any CO exiting the furnace will not burn at the combustion gap because the concentration will be below the lower flammability limit. If good heat transfer is not achieved in the furnace, some of the CO_2 will dissociate to CO and O_2 and as a result the CO concentration in the offgas will be higher than anticipated. If the combustion gap is not designed accordingly, there will be insufficient combustion air and some of the CO will report downstream to the gas cleaning equipment. If chlorine and metallic oxides (catalyst) are present in the offgas stream, there is a possibility of dioxin formation. The best way to ensure that this does not happen is to install a combustion chamber followed by a water spray quench to cool the gases below the temperature at which they will dissociate and react to form dioxins. This will also help to ensure low levels of CO downstream in the offgas system.

5.1.5.4 Conclusions

The following conclusions are drawn.

1. High levels of PCR in excess of 80% have been demonstrated for EAF operations.

2. An upper limit of 65% can be expected for HTE from post-combustion when there is cold scrap present. For post-combustion above the slag this will drop to 20%-30% when there is no scrap present. For post-combustion in the slag an HTE of 80% has been reported. A net efficiency of 50%-60% (PCR 3 HTE) may be achieved if DRI/HBI is added to absorb the energy released by post-combustion or if good heat transfer is achieved via circulation of metal droplets in the slag. During scrap melting operations, post-combustion above the slag should give the best results. During flat bath operations, it appears to be advantageous to burn the CO in or just above the slag.

3. Environmental benefits due to post-combustion have been demonstrated, but additional work needs to be carried out to better understand and optimize these benefits.

4. Potential gains due to post-combustion are highly dependent on the individual EAF operation efficiencies. Most applications of post-combustion in the EAF achieve an energy savings of 20-40kW · h/t.

5. The economic lower limit for CO level leaving the EAF needs to be established but will likely be in the range of 5%-10% based on environmental considerations.

6. If it is attempted to post-combust all of the CO in the EAF, yield losses and increased electrode consumption will occur.

7. Attempts to post-combust all of the CO in the furnace will likely have a negative effect on NO_x levels.

8. Oxygen injection into the bath should start early so that post-combustion can be carried out while the scrap is still relatively cold and is capable of absorbing the heat generated. In order to be most effective decarburization oxygen needs to be distributed throughout the bath. This will help to reduce local iron oxidation in the bath (and therefore the amount of EAF dust generated) and will also distribute the CO that is generated throughout the furnace which will help to maximize energy recovery once it is post-combusted.

9. Staged post-combustion similar to that carried out in the EOF preheat chambers has the greatest potential for capturing the energy generated through post-combustion. This is because the energy from the post-combustion reaction can be transferred to the scrap thus cooling the offgas and avoiding dissociation. A similar effect can probably be achieved in the shaft furnace. In the conventional EAF it will not be possible to recover as much of this energy, and some gas dissociation will likely take place.

10. Post-combustion in the slag can result in yield losses and an increase in the amount of EAF dust generated unless a thick slag layer is used. Alternatively partial postcombustion of the CO in the slag may prove to be more effective than complete post-combustion. The additional fluxes and energy consumed to provide a thick slag layer might offset any savings from post-combustion. In some operations where scrap preheating takes place, the dust is captured on the scrap and a conventional slag cover may be acceptable for post-combustion in the slag.

11. The optimum post-combustion strategy will vary from one operation to the next. Careful deliberation must be undertaken to determine the most cost effective means of applying post-combustion to EAF operations. Frequently, the optimum approach will involve selecting specific portions of the operating cycle in which post-combustion can be applied to give the highest returns. Other issues such as operability and maintenance requirements will also help to determine the complexity of the strategy employed.

Complete post-combustion of CO will be difficult to achieve and is likely uneconomical in light of possible detrimental effects (yield loss, refractory wear, electrode consumption, damage to furnace panels). Thus a complete analysis should be carried out on each installation to determine the best post-combustion practice for that location. Such a program should include CO, CO_2, H_2, H_2O, N_2, NO_x, SO_x gas analysis at the elbow, offgas flowrate and temperature exiting the EAF and at the baghouse, offgas analysis at the baghouse, analysis of baghouse dust before and after post-combustion for increases in iron content, monitoring of iron yield, electrode consumption and other consumables, monitoring EAF cooling water temperatures, slag analysis (check iron levels), and monitoring alloy consumption (check alloy yield).

Post-combustion can be an effective tool for the EAF operator but an economical operating practice must be established based on individual site criteria. There is no universal recipe which can apply for every facility. The way in which post-combustion can be applied is highly dependent on raw materials and operating practices and these must be thoroughly evaluated when implementing a post-combustion practice.

5.1.6 EAF Bottom Stirring

For conventional AC melting of scrap there is little natural convection within the bath. Temperature gradients have been reported in the range of 40-70℃. If there is limited bath movement, large pieces of scrap can take considerably longer to melt unless they are cut up as discussed previously under oxygen lancing operations. Concentration gradients within the bath can also lead to reduced reaction rates and over or under reaction of some portions of the bath.

The concept of stirring the bath is not a new one and records indicate that electromagnetic coils were used for stirring trials as early as 1933. Japanese studies indicated that flow velocities are much lower for stainless steels as compared to carbon steels for electromagnetic stirring. Studies indicate that electromagnetic stirring is capable of supplying sufficient stirring in some cases. However, the cost for electromagnetic stirring is high and it is difficult to retrofit into an existing operation.

Most EAF stirring operations presently in use employ gas as the stirring medium. These operations use contact or non-contact porous plugs to introduce the gas into the furnace. In some cases tuyeres are still used. The choice of gas used for stirring seems to be primarily argon or nitrogen though some trials with natural gas and with carbon dioxide have also been attempted. In a conventional EAF three plugs are located midway between the electrodes. For smaller furnaces a single plug centrally located appears to be sufficient. In EBT and other bottom tapping operations, the furnace tends to be elliptical and the nose of the furnace tends to be a cold spot. A stirring element is commonly located in this part of the furnace to promote mixing and aid in meltdown. Some operations have also found it beneficial to inject inert gas during tapping to help push the slag back and prevent slag entrainment in the tap stream. For common steel grades the gas flowrate for a contact system is typically 0.03-0.17Nm^3/min (1-6scfm) with a total consumption of 0.1-0.6Nm^3/t (3-20scf/t). Non-contact systems appear to use higher gas flow rates. Service life for contact systems is in the range of 300-500 heats. Some non-contact systems have demonstrated lives of more than 4000 heats. Fig. 5.14 shows several commercially available stirring elements.

Some of the main advantages attributed to bottom stirring include: reduction in carbon boils and cold bottoms, yield increases of 0.5%-1%, time savings of 1-16min (typical is 5min) per heat, energy savings of up to 43kW·h/t (typical is 10-20kW·h/t), improved alloy recovery, increased sulfur and phosphorus removal, and reduced electrode consumption.

In various operations surveyed the cost savings have ranged from $0.90 to $2.30 per ton.

5.1.7 Furnace Electrics

In addition to increased oxygen use in the EAF, considerable effort has been devoted towards maximizing electrical efficiency. This has been partly due to the fact that there are practical limitations

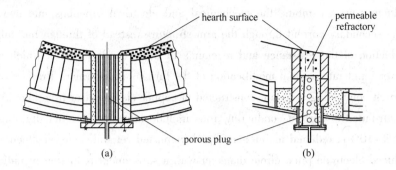
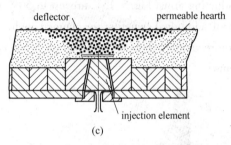

Fig. 5.14 Stirring element configurations (Courtesy of Center for Materials Production)
(a) Radex DPP (contact); (b) EF-KOA (non-contact); (c) Thyssen long-time stirrer (non-contact)

as to the amount of oxygen used in any one operation (due to environmental concerns). In addition, it has been realized that by using longer arc, lower current operations, it is possible to achieve much more efficient power input to the furnace. Several innovations have contributed to increased efficiencies in this area and are discussed in the following sections.

5.1.7.1 Electrode Regulation

In the days when furnaces melted small sized heats (5-20tons), electrode regulation was not a major concern. As furnaces became bigger and operating voltages increased however, it became necessary to control the electrodes more closely in order to maximize the efficiency of power input to the furnace. Over the past five years, there have been some substantial advances made in electrode regulation. These coupled with advances in furnace hydraulics (thus allowing for faster electrode response) have lead to considerable improvements in EAF operation. For one installation the following benefits were obtained following an upgrade of the electrode regulation system: power consumption decreased by 5%, flicker decreased by 10%, broken electrodes decreased by 90%, electrode consumption decreased by 8.5%, tap-to-tap time decreased by 18.5%, and average power input increased by 8.5%.

5.1.7.2 Current Conducting Arms

In conventional EAF design, the current is carried to the electrodes via bus tubes. These bus tubes usually contribute approximately 35% of the total reactance of the secondary electrical system.

Current conducting arms combine the mechanical and electrical functions into one unit. These arms carry the secondary current through the arm structure instead of through bus tubes. This results in a reduction of the resistance and reactance in the secondary circuit which allows an increase in power input rate without modification of the furnace transformer. Productivity is also increased. Current conducting arms are constructed of copper clad steel or from aluminum. Some of the benefits attributed to current conducting arms include increased productivity, increased power input rate (5%-10%), reduced maintenance and increased reliability, lower electrode consumption, and reduced electrode pitch circle diameter with a subsequent reduction in radiation to sidewalls.

The aluminum current conducting arms(Fig. 5. 15), are up to 50% lighter than conventional or copper clad steel arms. Several additional benefits are claimed including higher electrode speeds resulting in improved electrode regulation, reduced strain on electrode column components, and less mechanical wear of components.

Fig. 5. 15 Aluminum current conducting electrode arms (Courtesy of Badische Stahl Technology)

In comparison to the overall weight of arms and column, the weight of the arms should not be a significant factor.

5. 1. 8 High Voltage AC Operations

In recent years a number of EAF operations have retrofitted new electric power supplies in order to supply higher operating voltages. Energy losses in the secondary circuit are dependent on the secondary circuit reactance and to a greater extent on the secondary circuit current. If power can be supplied at a higher voltage, the current will be lower for the same power input rate. Operation with a lower secondary circuit current will also give lower electrode consumption. Thus it is advantageous to operate at as high a secondary voltage as is practical. Of course this is limited by arc flare to the sidewall and the existing furnace electrics. A good foamy slag practice can allow voltage increases of up to 100% without adversely affecting flare to the furnace sidewalls. Energy losses can be minimized when reactance is associated with the primary circuit.

Supplementary reactance is not a new technology. In the past, supplementary reactors were used to increase arc stability in small furnaces where there was insufficient secondary reactance. However, in the past few years this method has been used to increase the operating voltages on the EAF secondary circuit. This is achieved by connecting a reactor in series with the primary windings of the EAF transformer. This allows operation at a power factor of approximately $\sqrt{2}/2$ which is the theoretical optimum for maximum circuit power. This is made possible because the arc sees a large storage device in front of it in the circuit, which in effect acts as an electrical flywheel during operation. The insertion of the series reactor drops the secondary voltage to limit the amount of power transferred to the arc. In order to compensate for this, the furnace transformer secondary voltage is increased into the 900-1200V range allowing operation at higher arc voltages and lower electrode currents. Some of the benefits attributed to this type of operation are a more stable arc than for standard operations, electrode consumption reduced by 10%, secondary voltage increased by 60%-80%, power savings of 10-20kW · h/t, system power factor of approximately 0.72, furnace power factor of approximately 0.90, lower electrical losses due to lower operating current, and voltage flicker is reduced up to 40%.

5.1.9 DC EAF Operations

The progress in high power semiconductor switching technology brought into existence low cost efficient DC power supplies. Due to these advances, the high power DC furnace operation became possible. North American interest in DC furnace technology is growing with several existing installations and others that are currently under installation. The DC arc furnace is characterized by rectification of three phase furnace transformer voltages by thyristor controlled rectifiers. These devices are capable of continuously modulating and controlling the magnitude of the DC arc current in order to achieve steady operation. DC furnaces use only one graphite electrode with the return electrode integrated into the furnace bottom. There are several types of bottom electrodes: conductive hearth bottom, conductive pin bottom, single or multiple billet, and conductive fins in a monolithic magnesite hearth.

All of these bottom return electrode designs have been proven. The ones that appear to be used most often are the conductive pin bottom where a number of pins are attached to a plate and form the return path and the bottom billet design. The bottom electrode is air cooled in the case of the pin type and water-cooled in the case of the billet design. The area between pins is filled with ramming mix and the tip of the pins is at the same level as the inner furnace lining. As the refractory wears, the pins also melt back. DC furnaces operate with a hot heel in order to ensure an electrical path to the return electrode. During startup from cold conditions, a mixture of scrap and slag is used to provide an initial electrical path. Once this is melted in, the furnace can be charged with scrap.

Some of the early benefits achieved with DC operation included reduced electrode consumption (20% lower than high voltage AC, 50% lower than conventional AC), reduced voltage flicker (50%-60% of conventional AC operation) and reduced power consumption (5%-10% lower than for AC).

These results were mainly achieved on smaller furnaces which were retrofitted from AC to DC operation. However, some larger DC furnace installations did not immediately achieve the claimed benefits. Notably, two areas of concern emerged: electrode consumption and refractory consumption.

Several DC furnace operations found that the decrease in electrode consumption expected under DC operation did not occur. Much analysis by the electrode manufacturers indicated that physical conditions within the electrodes was different for AC and DC operations. As a result, for large DC electrodes carrying very large current, an increased amount of cracking and spalling was observed as compared to AC operations. Therefore, it was necessary to develop electrodes with physical properties better suited to DC operation. The economical maximum size for DC furnaces tends to be a function of limitations due to electrode size and current carrying capacity. At the present time the maximum economical size for a single graphite electrode DC furnace appears to be about 165 tons. Larger furnace sizes can be accommodated by using more than one graphite electrode. Fig. 5.16 shows furnace dimensions for a Kvaerner Clecim DC EAF.

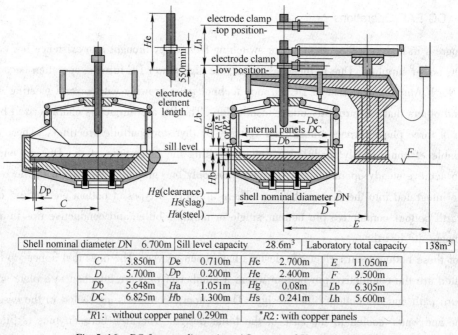

Fig. 5.16 DC furnace dimensions (Courtesy of Kvaerner Metals)

Several of the early DC operations experienced problems with refractory wear and bottom electrode life. These problems were directly related to arc flare within the furnace. The anode design has the greatest influence on the arc flare. In all DC furnaces, the electric arc is deflected in the direction opposite to the power supply due to assymetries in magnetic fields which are generated by the DC circuit. Thus the arc tends to concentrate on one area within the furnace creating a hot spot and resulting in excessive refractory wear. This is shown in Fig. 5.17. Several solutions have been developed to control or eliminate arc flare. All commercial bottom electrode designs are now configured to force the arc to the center of the furnace. In the case of bottom conductive refractory and the pin type bottom, it is necessary to provide split feed lines to the bottom anode or a bottom coil which helps to modify the net magnetic field generated. In the billet bottom design, the amount of current to each billet is controlled along with the direction of anode supply in order to control the arc. The bottom fin design utilizes the fact that electrical feed occurs at several points in order control arc deflection. Quadrants located further from the rectifier are supplied with higher current than those located closer to the rectifier.

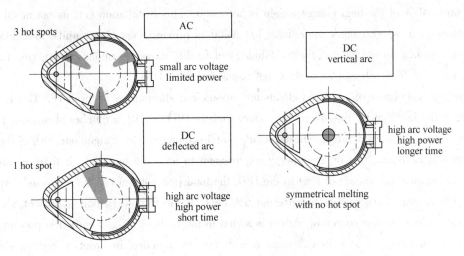

Fig. 5.17 Electric arc behavior (Courtesy of Kvaerner Metals)

Some feel that the possibility for increased automation of EAF activities is greater for the DC furnace. This is because with only one electrode, there is increased space both on top and within the furnace. DC furnace installations can be expected to cost from 10%-35% more than a comparable AC installation. However, calculations on payback indicate that this additional cost can be recovered in one to two years due to lower operating costs.

Bowman conducted an analysis comparing AC and DC furnace operations and found that the electrical losses amount to approximately 4% in AC operations and 5.5% in DC operations; the difference in absolute terms is relatively insignificant. The difference in total energy consumption between AC and DC furnaces is likely less than 9kW · h/t in favor of the DC furnace, however many other variables influence the power consumption and it is difficult to develop accurate figures. DC furnaces experience roughly 25% less electrode consumption than AC furnaces, this correlating to typically 0.4kg/t. This difference appears to be greater for smaller AC furnaces. Flicker is approximately 60% lower for DC operations, however, advances in AC power system configurations (additional reactance) may reduce this difference to 40%.

Some typical results which have been presented for large DC EAF operations are electrode consumption of 1-2kg/t liquid steel, power consumption at 350-500kW · h/t liquid steel, tap-to-tap times of 45-120minutes, and bottom life of 1500-4000 heats. It is important to remember however, that power consumption is highly dependent on operating practices, tap temperature, use of auxiliary fuels, scrap type etc.

5.1.10 Use of Alternative Iron Sources in the EAF

Hot metal production is a standard part of operations in integrated steelmaking. Hot metal is produced in the blast furnace from iron ore pellets. This hot metal is then refined in basic oxygen furnaces to produce steel. However, several operations which were previously integrated operations are now charging hot metal to the EAF. One such installation is Cockerill Sambre in Belgium where up to 40% of the total charge weight is hot metal. This installation gets its hot metal from a blast furnace. In several other operations, hot metal is provided via Corex units, mini blast furnaces, or cupolas. In the case of the Saldahna Steel facility currently under construction in South Africa, the EAF feed will consist of 45% hot metal and 55% DRI.

There is a wide range of tabulated effects for various iron alternatives in the EAF. This is primarily due to the fact that within any given product such as DRI or HBI, a number of process parameters may vary quite considerably. These include metallization, carbon, gangue, etc. All of these parameters will have an affect on the energy requirement to melt the material. If there is sufficient carbon to balance the amount of FeO in the DRI, the total iron content can be recovered. Approximately 1% carbon is required to balance out 6% FeO. If insufficient carbon is present, yield loss will result unless another source of carbon is added to the bath. If excess carbon is present, it can be used as an energy source in conjunction with oxygen injection in order to reduce electrical power requirements. Generally speaking, DRI requires 100-200 additional kW·h/t to melt as compared to scrap melting. If up to 25% DRI is to be used in the charge makeup, it can be added in the bucket. If a larger percentage is to be used it can be fed continuously through the roof. One advantage of DRI is that it can be fed continuously with power on and therefore no thermal losses are incurred by opening the roof. HBI is more dense than DRI and as a result can be charged in the scrap bucket without increasing the number of charges required. DRI tends to float at the slag bath interface while HBI, which has a much higher density, tends to sink into the bath and melt in a manner similar to pig iron. The amount of silica present in the DRI will have a large effect on the economics of steel production from DRI. Silica will attack refractories unless sufficient lime is present to neutralize its effect. In general, a V-ratio of 2.5-3.0 is desired for good slag foaming. Thus the lime requirement increases greatly if silica levels in the DRI are high. Melting power requirements increase accordingly.

In the case of cold pig iron, a power savings should result from using 10%-15% pig iron in the charge. This is due to the silicon and carbon contained in the pig iron. These act as a source of chemical heat in the bath when oxygen is injected. Pig iron typically contains up to 0.65% silicon which reacts with oxygen to produce silica which reports to the slag. This requires some additional lime addition in order to maintain the slag basicity. Usually, a maximum of 20% cold pig iron is used in the EAF because it takes longer to melt in than scrap, especially if it is supplied in large pieces. Small sized pieces are preferable. The pig iron can contain up to 4% carbon which results in a very high bath carbon level. Removal of this carbon with oxygen generates much heat but also requires increased blowing times because practical limits exist on the rate at which oxygen can be blown into the steel.

Iron carbide can be charged into the furnace in sacks or it can be injected. Injection is the preferred method of introducing the material into the bath as recovery is maximized. This however creates some practical limitations as to the quantity of iron carbide used since limitations on the injection rate exist. At Nucor Crawfordsville, the maximum rate achieved so far is 2500kg/min. Iron carbide dissolves into the steel bath and as the carbon goes into solution, it reacts with FeO or dissolved oxygen in the bath producing a very fine dispersion of carbon monoxide bubbles. These bubbles are very beneficial because they help to strip nitrogen from the bath. Depending on the degree of metallization in the iron carbide, the energy requirements for dissolution of the iron carbide also vary.

The charging of hot metal to the EAF sounds like a simple proposition though it is in fact quite complex. Care must be taken that the hot metal which is charged does not react with the highly oxidized slag which is still in the EAF. Some operations charge hot metal to the EAF by swinging the roof and pouring it into the furnace. This causes very rapid mixing of the hot heel and the highly oxidized slag in the EAF with the hot metal and sometimes explosions do occur. Thus for this mode of operation it is recommended that a slag deoxidizer be added prior to hot metal addition. Typical deoxidizers are silicon fines, aluminum fines and calcium carbide. An alternative method of charging the hot metal to the EAF is to pour it down a launder which is inserted into the side of the EAF. This method requires more time for charging of the hot metal but results in a much safer operation. Paul Wurth has recently developed a side charging system whereby the hot metal can be charged while power is on and thus the charging time is not an issue.

5.1.11 Conclusions

It is apparent that there are many technologies available for improving EAF operating efficiency. The general results for these have been listed in an attempt to provide a starting point for those in the process of upgrading operations. Of course results will vary from one installation to the next. However, if a conservative approach is taken and the median of the reported results is used for calculation purposes, the results can be expected to be achievable. It is important to evaluate the effects of certain processes both on furnace operations and on other systems such as fume control. The use of substitute fuels (oxygen and natural gas) may be limited by the capacity of the fume system. If upgrades are to be made, one must also evaluate the need for upgrades to auxiliary systems. The data presented in this section provides a starting point for the person evaluating process changes to improve EAF efficiency. The technologies that have been reviewed are well proven. The interaction between these processes has not been evaluated, though in some cases the blending of these operations can prove to be most beneficial (eg. oxygen lancing, slag foaming and CO postcombustion). When evaluating upgrade requirements for a particular operation, it is necessary to clearly list the objectives and then match these with suitable technologies. It is important to maintain a global perspective regarding overall costs and operations in order to arrive at true optimal operating efficiency in the EAF.

5.2 Raw Materials

The main raw material for EAF steelmaking is steel scrap. Scrap is an energy intensive and valuable commodity and comes primarily from three main sources: reclaimed scrap (also known as obsolete scrap) which is obtained from old cars, demolished buildings, discarded machinery and domestic objects; industrial scrap (also known as prompt scrap) which is generated by industries using steel within their manufacturing processes; and revert scrap (also known as home scrap) which is generated within the steelmaking and forming processes (e.g. crop ends from rolling operations, metallic losses in slag etc.).

The latter two forms of scrap tend to be clean, i.e. they are close in chemical composition to the desired molten steel composition and thus are ideal for recycle. Reclaimed/obsolete scrap frequently has a quite variable composition and quite often contains contaminants that are undesirable for steelmaking. Levels of residual elements such as Cu, Sn, Ni, Cr, and Mo are high in obsolete scrap and can affect casting operations and product quality if they are not diluted. Thus a facility which has a need for very low residual levels in the steel will be forced to use higher quality prompt scrap but at a much higher cost. The alternative is to use a combination of the contaminated obsolete scrap along with what are generally referred to as clean iron units or virgin iron units. These are materials which contain little or no residual elements. Clean iron units are typically in the form of direct reduced iron (DRI), hot briquetted iron (HBI), iron carbide, pig iron, and molten pig iron (hot metal).

It is possible to use lower grade scrap which contains residual elements, if this scrap is blended with clean iron units so that the resulting residual levels in the steel following melting meet the requirements for flat rolled products.

Obsolete scrap is much more readily available than prompt scrap and thus the use of clean iron units is expected to increase as shortages of prompt scrap continue to grow.

In addition to classification of scrap into the above three groups, scrap is also classified based on its physical size, its source and the way in which it is prepared. For example the following categories are those commonly used in North America: No. 1 bundles, No. 1 factory bundles, No. 1 shredded, No. 1 heavy melt, No. 2 heavy melt, No. 2 bundles, No. 2 shredded, busheling, turnings, shredded auto, structural/plate 3ft(0.9m), structural/plate 5ft(1.5m), rail crops, and rail wheels.

In addition to the residual elements contained in the scrap, there are also several other undesirable components including, oil, grease, paint coatings, zinc coatings, water, oxidized material and dirt. The lower the grade of scrap, the more likely it is to contain greater quantities of these materials. As a result this scrap may sell at a discount but the yield of liquid steel may be considerably lower than that obtained when using a higher grade scrap. In addition, these undesirable components may result in higher energy requirements and environmental problems. Thus the decision for scrap mix to be used within a particular operation will frequently depend on several factors including availability, scrap cost, melting cost, yield, and the effect on operations (based on scrap density, oil and grease content, etc.).

In practice, most operations buy several different types of scrap and blend them to yield the most desirable effects for EAF operations.

5.3 Fluxes and Additives

Carbon is essential to the manufacture of steel. Carbon is one of the key elements which give various steel grades their properties. Carbon is also important in steelmaking refining operations and can contribute a sizable quantity of the energy required in steelmaking operations. In BOF steelmaking, carbon is present in the hot metal that is produced in the blast furnace. In electric furnace steelmaking, some carbon will be contained in the scrap feed, in DRI, HBI or other alternative iron furnace feeds. The amount of carbon contained in these EAF feeds will generally be considerably lower than that contained in hot metal and typically, some additional carbon is charged to the EAF. In the past carbon was charged to the furnace to ensure that the melt-in carbon level was above that desired in the final product. As higher oxygen utilization has developed as standard EAF practice, more carbon is required in EAF operations. The reaction of carbon with oxygen within the bath to produce carbon monoxide results in a significant energy input to the process and has lead to substantial reductions in electrical power consumption in EAF operations. The generation of CO within the bath is also the key to achieving low concentrations of dissolved gases (nitrogen and hydrogen) in the steel as these are flushed out with the carbon monoxide. In addition, oxide inclusions are flushed from the steel into the slag.

In oxygen injection operations, some iron is oxidized and reports to the slag. Oxy-fuel burner operations will also result in some scrap oxidation and this too will report to the slag once the scrap melts in. Dissolved carbon in the steel will react with FeO at the slag/bath interface to produce CO and recover iron units to the bath.

The amount of charge carbon used will be dependent on several factors including carbon content of scrap feed, projected oxygen consumption, desired tap carbon, and the economics of iron yield versus carbon cost. In general, the amount used will correspond to a carbon/oxygen balance as the steelmaker will try to maximize the iron yield. Typical charge carbon rates for medium carbon steel production lie in the range of 2-12kg/t of liquid steel.

Generally, the three types of carbonaceous material used as charge carbon in EAF operations are anthracite coal, metallurgical coke and green petroleum coke. Most anthracite coal used in North American steelmaking operations is mined in eastern Pennsylvania. This material has a general composition of 3%-8% moisture content, 11%-18% ash content and 0.4%-0.7% sulfur.

The high variation in ash content translates into wide variations in fixed carbon content and in general, EAF operations strive to keep ash content to a minimum. The ash consists primarily of silica. Thus increased ash input to the EAF will require additional lime addition in order to maintain the desired V-ratio in the slag. The best grades of anthracite coal have fixed carbon contents of 87%-89%. Low grade anthracite coals may have fixed carbon levels as low as 50%.

Anthracite coal is available in a wide variety of sizes ranging from 4in × 8in (10.16cm × 20.32cm) down to 3/64in(7.62/162.56cm) ×100 mesh(150μm). The most popular sizes for use as charge carbon are nut [5/8in×13/16in(12.70/20.32cm×33.02/40.64cm)], pea [13/16in× 9/16in(33.02/40.64cm×22.86/40.64cm)], and buckwheat [9/16in×5/16in(22.86/40.64cm× 12.70/40.64cm)].

Metallurgical coke is produced primarily in integrated steel operations and is used in the blast furnace. However, some coke is used as EAF charge carbon. Generally, this material has a composition of 1%-2% moisture content, 1%-3.5% volatile material, 86%-88% fixed carbon, 9%-12% ash content and 0.88%-1.2% sulfur.

Usually coke breeze with a size of -1/2×0 is used as charge carbon. Coarser material can be used but is more expensive.

Green petroleum coke is a byproduct of crude oil processing. Its properties and composition vary considerably and are dependent on the crude oil feedstock from which it is derived. Several coking processes are used in commercial operation and these will produce considerably different types of coke.

Sponge coke results from delayed coker operations and is porous in nature. It may be used as a fuel or may be processed into electrodes or anodes depending on sulfur content and impurity levels. This material is sometimes available as charge carbon.

Needle coke is produced using a special application of the delayed coker process. It is made from high grade feedstocks and is the prime ingredient for the production of carbon and graphite electrodes. This material is generally too expensive to be used as charge carbon.

Shot coke is a hard, pebble-like material resulting from delayed coker operation under conditions which minimize coke byproduct generation. It is generally used as a fuel and is cost competitive as charge carbon.

Fluid coke is produced in a fluid coker by spraying the residue onto hot coke particles. It is usually high in sulfur content and is used in anode baking furnaces. It can also be used as a recarburizer if it is calcined.

Over the past decade, many operations have adopted foamy slag practices. At the start of meltdown the radiation from the arc to the sidewalls is negligible while the electrodes are surrounded by the scrap. As melting proceeds the efficiency of heat transfer to the scrap and bath drops off and more heat is radiated from the arc to the sidewalls. By covering the arc in a layer of slag, the arc is shielded and the energy is transferred to the bath. Oxygen is injected with coal to foam the slag by producing CO gas in the slag. In some cases only carbon is injected and the carbon reacts with FeO in the slag to produce CO gas. When foamed, the slag cover increases from 10 to 30 cm thick. In some cases the slag is foamed to such an extent that it comes out of the electrode ports. Claims for the increase in energy transfer efficiency range from an efficiency of 60%-90% with slag foaming compared to 40% without. It has been reported that at least 0.3% carbon should be removed from the bath using oxygen in order to achieve a good foamy slag practice.

The effectiveness of slag foaming is dependent on several process parameters as described in the section on furnace technologies. Typical carbon injection rates for slag foaming are 2-5kg/t of liquid steel for low to medium powered furnaces. Higher powered furnaces and DC furnaces will tend to use 5-10kg of carbon per ton liquid steel. This is due to the fact that the arc length is much greater than low powered AC operations and therefore greater slag cover is required to bury the arc.

Lime is the most common flux used in modern EAF operations. Most operations now use basic refractories and as a result, the steelmaker must maintain a basic slag in the furnace in order to minimize refractory consumption. Slag basicity has also been shown to have a major effect on slag foaming capabilities. Thus lime tends to be added both in the charge and also via injection directly into the furnace. Lime addition practices can vary greatly due to variances in scrap composition. As elements in the bath are oxidized (e.g. P, Al, Si, Mn) they contribute acidic components to the slag. Thus basic slag components must be added to offset these acidic contributions. If silica levels in the slag are allowed to get too high, significant refractory erosion will result. In addition, FeO levels in the slag will increase because FeO has greater solubility in higher silica slags. This can lead to higher yield losses in the EAF.

The generation of slag also allows these materials which have been stripped from the bath to be removed from the steel by pouring slag out of the furnace through the slag door which is located at the back of the EAF. This is known as slagging off. If the slag is not removed but is instead allowed to carry over to the ladle it is possible for slag reversion to take place. This occurs when metallic oxides are reduced out of the slag by a more reactive metallic present in the steel. When steel is tapped it is frequently killed by adding either silicon of aluminum during tapping. The purpose of these additions is to lower the oxygen content in the steel. If however P_2O_5 is carried over into the ladle, it is possible that it will react with the alloy additions producing silica or alumina and phosphorus which will go back into solution in the steel.

Sometimes magnesium lime is added to the furnace either purely as MgO or as a mixture of MgO and CaO. Basic refractories are predominantly MgO, thus by adding a small amount of MgO to the furnace, the slag can quickly become saturated with MgO and thus less refractory erosion is likely to take place.

5.4 Furnace Operations

5.4.1 EAF Operating Cycle

The electric arc furnace operates as a batch process. Each batch of steel that is produced is known as a heat. The electric arc furnace operating cycle is known as the tap-to-tap cycle. The tap-to-tap cycle is made up of the following operations: furnace charging, melting, refining, de-slagging, tapping and furnace turnaround. Modern operations aim for a tap-to-tap cycle of less than 60 minutes. With the advance of EAF steelmaking into the flat products arena, tap-to-tap times of 35-40 minutes are now being sought with twin shell furnace operations.

A typical 60 minutes tap-to-tap cycle is:

first	charge 3 minutes
first	meltdown 20 minutes
second	charge 3 minutes
second	meltdown 14 minutes
refining	10 minutes
tapping	3 minutes
turnaround	7 minutes
Total	60 minutes

5.4.2 Furnace Charging

The first step in the production of any heat is to select the grade of steel to be made. Usually a heat schedule is developed prior to each production shift. Thus the melter will know in advance the schedule for the shift. The scrap yard operators will batch buckets of scrap according to the needs of the melter. Preparation of the charge bucket is an important operation, not only to ensure proper melt-in chemistry but also to ensure good melting conditions. The scrap must be layered in the bucket according to size and density in order to ensure rapid formation of a liquid pool in the hearth while also providing protection of the sidewalls and roof from arc radiation. Other considerations include minimization of scrap cave-ins which can break electrodes and ensuring that large heavy pieces of scrap do not lie directly in front of burner ports which would result in blow-back of the flame onto the water-cooled panels. The charge can include lime and carbon or these can be injected into the furnace during the heat. Many operations add some lime and carbon in the scrap bucket and supplement this with injection.

The first step in any tap-to-tap cycle is charging of the scrap. The roof and electrodes are raised and are swung out to the side of the furnace to allow the scrap charging crane to move a full bucket of scrap into place over the furnace. The bucket bottom is usually a clam shell design—i.e. the bucket opens up by retracting two segments on the bottom of the bucket, see Fig. 5.18. Another common configuration is the "orange peel" design. The scrap falls into the furnace and the scrap crane removes the scrap bucket. The roof and electrodes swing back into place over the furnace. The roof is lowered and then the electrodes are lowered to strike an arc on the scrap. This commences the melting portion of the cycle. The number of charge buckets of scrap required to produce a heat of steel is dependent primarily on the volume of the furnace and the scrap density. Most modern furnaces are designed to operate with a minimum of back charges. This is advantageous because charging is dead time, whereby the furnace does not have power on and therefore is not melting. Minimizing these dead times helps to maximize the productivity of the furnace. In addition, energy is lost each time the furnace roof is opened. This can amount to 10-20kW · h/t for each occurrence. Most operations aim for 2-3 buckets of scrap per heat and will attempt to blend their scrap to meet this requirement. Some operations achieve a single bucket charge. Continuous charging operations such as Consteel and the Fuchs shaft furnace eliminate the charging cycle.

Fig. 5.18 Clam shell bucket charging scrap to the EAF (Courtesy of SMS GHH)

5.4.3 Melting

The melting period is the heart of EAF operations. The EAF has evolved into a highly efficient melting apparatus and modern designs are focused on maximizing its melting capacity. Melting is accomplished by supplying energy to the furnace interior. This energy can be electrical or chemical. Electrical energy is supplied via the graphite electrodes and is usually the largest contributor in melting operations. Initially, an intermediate voltage tap is selected until the electrodes can bore into the scrap. Usually, light scrap is placed on top of the charge to accelerate bore-in. After a few minutes, the electrodes will have penetrated the scrap sufficiently that a long arc (high voltage) tap can be used without fear of radiation damage to the roof. The long arc maximizes the transfer of power to the scrap and a liquid pool of metal will form in the furnace hearth. Approximately 15% of the scrap is melted during the initial bore-in period. At the start of melting the arc is erratic and unstable. Wide swings in current are observed accompanied by rapid movement of the electrodes. As the furnace atmosphere heats up the arcing tends to stabilize and once the molten pool is formed, the arc becomes quite stable and the average power input increases.

Chemical energy can be supplied via several sources such as oxy-fuel burners and oxygen lancing. Oxy-fuel burners burn natural gas using oxygen or a blend of oxygen and air. Heat is transferred to the scrap by radiation and convection. Heat is transferred within the scrap by conduction. In some operations, oxygen is used to cut scrap. Large pieces of scrap take longer to melt into the bath than smaller pieces. A consumable pipe lance can be used to cut the scrap. The oxygen reacts with the hot scrap and burns iron to produce intense heat for cutting the scrap. Once a molten pool of steel is generated in the furnace, oxygen can be lanced directly into the bath. This oxygen will react with several components in the bath including, aluminum, silicon, manganese, phosphorus, carbon and iron. All of these reactions are exothermic (i. e. they generate heat) and will supply energy to aid in the melting of the scrap. The metallic oxides which are formed will eventually reside in the slag. The reaction of oxygen with carbon in the bath will produce carbon monoxide which may burn in the furnace if there is oxygen available. Otherwise the carbon monoxide will carry over to the direct evacuation system.

Once enough scrap has been melted to accommodate the second charge, the charging process is repeated. After the final scrap charge is melted, the furnace sidewalls can be exposed to high radiation from the arc. As a result, the voltage must be reduced. Alternatively, creation of a foamy slag will allow the arc to be buried and will protect the furnace shell. In addition, a greater amount of energy will be retained in the slag and is transferred to the bath resulting in greater energy efficiency. When the final scrap charge is fully melted, flat bath conditions are reached. At this point, a bath temperature and a chemical analysis sample will be taken. The analysis of the bath chemistry will allow the melter to determine the amount of oxygen to be blown during refining. The melter can also start to arrange for the bulk tap alloy additions to be made. These quantities are confirmed following refining.

5. 4. 4 Refining

Refining operations in the electric arc furnace have traditionally involved the removal of phosphorus, sulfur, aluminum, silicon, manganese and carbon. In recent times, dissolved gases in the bath have also become a concern, especially nitrogen and hydrogen levels. Traditionally, refining operations were carried out following meltdown , i. e. once a flat bath was achieved. These refining reactions are all dependent upon oxygen being available. Oxygen was lanced at the end of meltdown to lower the bath carbon content to the desired level for tapping. Most of the compounds which are to be removed during refining have a higher affinity for oxygen than carbon. Thus the oxygen will preferentially react with these elements to form oxides which will report to the slag.

In modern EAF operations, especially those operating with a hot heel, oxygen may be blown into the bath throughout the heat cycle. As a result, some of the refining operations occur concurrent with melting.

Most impurities such as phosphorous, sulfur, silicon, aluminum and chromium are partially removed by transfer to the slag. In particular the equilibrium partition ratio between metal and slag are given as functions of slag chemistry and temperature.

The slag in an EAF operation will, in general, have a lower basicity than that for oxygen steelmaking. In addition, the quantity of slag per ton of steel will also be lower in the EAF. Therefore the removal of impurities in the EAF is limited. A typical slag composition is presented in Table 5. 4.

Table 5.4 Typical slag constituents

Component	Source	Composition range/%
CaO	Charged	40-60
SiO$_2$	Oxidation product	5-15
FeO	Oxidation product	10-30
MgO	Charged as dolomite	3-8
CaF$_2$	Charged slag fluidizer	
MnO	Oxidation product	2-5
S	Absorbed from steel	
P	Oxidation product	

Once these materials enter into the slag phase they will not necessarily stay there. Phosphorus retention in the slag is a function of the bath temperature, the slag basicity and FeO levels in the slag. At higher temperature or low FeO levels, the phosphorus will revert from the slag back into the bath. Phosphorus removal is usually carried out as early as possible in the heat. Hot heel practice is very beneficial for phosphorus removal because oxygen can be lanced into the bath while the bath temperature is quite low. Early in the heat the slag will contain high FeO levels carried over from the previous heat. This will also aid in phosphorus removal. High slag basicity (i.e. high lime content) is also beneficial for phosphorus removal but care must be taken not to saturate the slag with lime. This will lead to an increase in slag viscosity which will make the slag less effective for phosphorus removal. Sometimes fluorspar is added to help fluidize the slag. Gas stirring is also beneficial because it will renew the slag/metal interface which will improve the reaction kinetics.

In general, if low phosphorus levels are a requirement for a particular steel grade, the scrap is selected to give a low level at melt-in. The partition ratio of phosphorus in the slag to phosphorus in the bath ranges from 5.0-15.0. Usually the phosphorus is reduced by 20%-50% in the EAF. However, the phosphorous in the scrap is low compared to hot metal and therefore this level of removal is acceptable. For oxygen steelmaking higher slag basicity and FeO levels give a partition ratio of 100 and with greater slag weight up to 90% of the phosphorous is removed.

Sulfur is removed mainly as a sulfide dissolved in the slag. The sulfur partition between the slag and metal is dependent on the chemical analysis and temperature of the slag (high basicity is better, low FeO content is better), slag fluidity (high fluidity is better), the oxidation level of the steel (which should be as low as possible), and the bath composition. Generally the partition ratio is 3.0-5.0 for EAF operations.

It can be seen that removal of sulfur in the EAF will be difficult especially given modern practices where the oxidation level of the bath is quite high. If high lime content is to be achieved in the slag, it may be necessary to add fluxing agents to keep the slag fluid. Usually the meltdown slag must be removed and a second slag built. Most operations have found it to be more effective to carry out desulfurization during the reducing phase of steelmaking. This means that desulfurization is commonly carried out during tapping (where a calcium aluminate slag is built) and during ladle furnace operations. For reducing conditions where the bath has a much lower oxygen activity, distribution ratios for sulfur of 20-100 can be achieved.

Control of the metallic constituents in the bath is important as it determines the properties of the final product. Usually, the melter will aim for lower levels in the bath than are specified for the final product. Oxygen reacts with aluminum, silicon and manganese to form metallic oxides which are slag components. These metallics tend to react before the carbon in the bath begins to react with the oxygen. These metallics will also react with FeO resulting in a recovery of iron units to the bath. For example:

$$Mn + FeO \Longleftrightarrow MnO + Fe \tag{5.6}$$

Manganese will typically be lowered to about 0.06% in the bath.

The reaction of carbon with oxygen in the bath to produce CO is important as it supplies energy to the bath and also carries out several important refining reactions at the same time. In modern EAF operations, the combination of oxygen with carbon can supply between 30% and 40% of the net heat input to the furnace. Evolution of carbon monoxide is very important for slag foaming. Coupled with a basic slag, CO bubbles will help to inflate the slag which will help to submerge the arc. This gives greatly improved thermal efficiency and allows the furnace to operate at high arc voltages even after a flat bath is reached. Submerging the arc helps to prevent nitrogen from being exposed to the arc where it will dissociate and become dissolved in the steel.

If the CO is evolved within the bath, it will also remove nitrogen and hydrogen from the steel. The capacity for nitrogen removal is dependent on the amount of CO generated in the metal. Nitrogen levels as low as 50ppm can be achieved in the furnace prior to tap. Bottom tapping is beneficial for maintaining low nitrogen levels as tapping is fast and a tight tap stream is maintained. A high oxygen content in the steel is beneficial for reducing nitrogen pickup at tap as compared to deoxidation of the steel at tap.

At 1600℃, the maximum solubility of nitrogen in pure iron is 450ppm. Typically, the nitrogen levels in the steel following tapping are 80-100ppm.

Decarburization is also beneficial for the removal of hydrogen. It has been demonstrated that decarburizing at a rate of 1% per hour can lower hydrogen levels in the steel from 8ppm down to 2ppm in ten minutes.

At the end of refining, a bath temperature measurement and a bath sample are taken. If the temperature is too low, power may be applied to the bath. This is not a big concern in modern meltshops where temperature adjustment is carried out in the ladle furnace.

5.4.5 Deslagging

Deslagging operations are carried out to remove impurities from the furnace. During melting and refining operations, some of the undesirable materials within the bath are oxidized and enter the slag phase.

Thus it is advantageous to remove as much phosphorus into the slag as early in the heat as possible (i.e. while the bath temperature is still low). Then the slag is poured out of the furnace through the slag door. Removal of the slag eliminates the possibility of phosphorus reversion.

During slag foaming operations, carbon may be injected into the slag where it will reduce FeO to metallic iron and will generate carbon monoxide which helps to inflate the slag. If the high phosphorus slag has not been removed prior to this operation, phosphorus reversion will occur.

5.4.6 Tapping

Once the desired bath composition and temperature are achieved in the furnace, the taphole is opened and the furnace is tilted so that the steel can be poured into a ladle for transfer to the next batch operation (usually a ladle furnace or ladle station). During the tapping process bulk alloy additions are made based on the bath analysis and the desired steel grade. Deoxidizers may be added to the steel to lower the oxygen content prior to further processing. This is commonly referred to as blocking the heat or killing the steel. Common deoxidizers are aluminum or silicon in the form of ferrosilicon or silicomanganese. Most carbon steel operations aim for minimal slag carryover. A new slag cover is built during tapping. For ladle furnace operations, a calcium laminate slag is a good choice for sulfur control. Slag forming compounds are added in the ladle at tap so that a slag cover is formed prior to transfer to the ladle furnace. Additional slag materials may be added at the ladle furnace if the slag cover is insufficient. Fig. 5.19 shows an EBT furnace tapping into a ladle.

Fig. 5.19 EBT furnace during tapping (Courtesy of Fuchs)

5.4.7 Furnace Turnaround

Furnace turnaround is the period following completion of tapping until the first scrap charge is dropped in the furnace for the next heat. During this period, the electrodes and roof are raised and the furnace lining is inspected for refractory damage. If necessary, repairs are made to the hearth, slagline, taphole and spout. In the case of a bottom tapping furnace, the taphole is filled with sand. Repairs to the furnace are made using gunned refractories or mud slingers. In most modern furnaces, the increased use of water-cooled panels has reduced the amount of patching or fettling required between heats. Many operations now switch out the furnace bottom on a regular basis (every 2-6 weeks) and perform the maintenance off-line. This reduces the power-off time for the EAF and maximizes furnace productivity. Furnace turnaround time is generally the largest dead time in the tap-to-tap cycle. With advances in furnace practices this has been reduced from 20 minutes to less than 5 minutes in some newer operations.

5.4.8 Furnace Heat Balance

To melt steel scrap, it takes a theoretical minimum of 300kW · h/t. To provide superheat above the melting point of 1520℃ (2768℉) requires additional energy and for typical tap temperature requirements, the total theoretical energy required usually lies in the range of 350-370kW · h/t. However, EAF steelmaking is only 55%-65% efficient and as a result the total equivalent energy input is usually in the range of 560-680kW · h/t for most modern operations. This energy can be supplied from a number of sources including electricity, oxy-fuel burners and chemical bath reactions. The typical distribution is 60%-65%, 5%-10% and 30%-40% respectively. The distribution selection will be highly dependent on local material and consumable costs and tends to be unique to the specific meltshop operation. A typical balance for both older and more modern EAFs is given in the Table 5.5.

Table 5.5 Typical energy balance for EAFs (%)

	Items	UHP Furnace	Low to medium power furnace
Inputs	Electrical energy	50-60	75-85
	Burners	5-10	
	Chemical reactions	30-40	15-25
	Total inputs	100	100
Outputs	Steel	55-60	50-55
	Slag	8-10	8-12
	Cooling water	8-10	5-6
	Miscellaneous	1-3	17-30
	Offgas	17-28	7-10
	Total outputs	100	100

Several factors are immediately apparent from these balances. Much more chemical energy is being employed in the EAF and correspondingly, electrical power consumption has been reduced. Furnace efficiency has improved with UHP operation as indicated by the greater percentage of energy being retained in the steel. Losses to cooling water are higher in UHP operation due to the greater use of water-cooled panels. Miscellaneous losses such as electrical inefficiencies were much greater for older, low powered operations. Energy loss to the furnace offgas is much greater in UHP furnace operation due to greater rates of power input and shorter tap-to-tap times.

Of course the figures in Table 5.5 are highly dependent on the individual operation and can vary considerably from one facility to another. Factors such as raw material composition, power input rates and operating practices (e.g. post-combustion, scrap preheating) can greatly alter the energy balance. In operations utilizing a large amount of charge carbon or high carbon feed materials, up to 60% of the energy contained in the offgas may be calorific due to large quantities of uncombusted carbon monoxide. Recovery of this energy in the EAF could increase energy input by 8%-10%. Thus it is important to consider such factors when evaluating the energy balance for a given furnace operation.

The International Iron and Steel Institute (IISI) classifies EAFs based on the power supplied per ton of furnace capacity. For most modern operations, the design would allow for at least 500kV · A/t of capacity. The IISI report on electric furnaces indicates that most new installations allow for 900-1000kV · A/t of furnace capacity. Most furnaces operate at a maximum power factor of 0.85. Thus the above transformer ratings would correspond to a maximum power input of 0.75-0.85MW/t of furnace capacity.

5.5 New Scrap Melting Processes

Over the past twenty years the steelmaking world has seen many changes in operating practices and the utilization of new process concepts in an attempt to lower operating costs and to improve product quality. In addition many new alternatives have been presented as lower cost alternatives to conventional AC EAF melting. Some of the specific objectives of these processes include lowering specific capital costs, increasing productivity, and improving process flexibility.

All of these processes share one or more of the following features in common. Energy from the waste offgas is used to preheat the scrap. Carbon is added to the bath and is later removed by oxygen injection in order to supply energy to the process. An attempt is made to combust the CO generated in the process to maximize energy recovery. An attempt is made to maximize power-on time and minimize turnaround time.

Many of these process have their roots associated with scrap preheating. It is only fitting that this be discussed first in order to provide the groundwork for the development of these processes.

5.5.1 Scrap Preheating

Scrap preheating has been used for over 30 years to offset electrical steel melting requirements primarily in regions with high electricity costs such as Japan and Europe. Scrap preheating involves the use of hot gas to heat scrap in the bucket prior to charging. The source of the hot gases can be either offgases from the EAF or gases from a burner. The primary energy requirement for the EAF is for heating of the scrap to its melting point. Thus energy can be saved if scrap is charged to the furnace hot. Preheating of scrap also eliminates the possibility of charging wet scrap which eliminates the possibility of furnace explosions. Scrap preheating can reduce electrical consumption and increase EAF productivity.

Some suppliers have noted that there is a maximum preheat temperature beyond which further efforts to heat the scrap lead to diminished returns. This temperature lies in the range of 540-650°C. It is estimated that by preheating the scrap to a temperature of 425-540°C, a total of 63-72kW · h/t of electrical energy can be saved. Early scrap preheaters used independent heat sources. The scrap was usually heated in the scrap bucket. Energy savings reported from this type of preheating were as high as 40kW · h/t with associated reductions in electrode and refractory consumption due to reduced tap-to-tap times.

As fourth hole offgas systems were developed, attempts were made to use the EAF offgas for scrap preheating. A side benefit that was reported was that the amount of baghouse dust decreased because the dust was sticking to the scrap during preheating. Scrap preheating with furnace offgas is difficult to control due to the variation in offgas temperature throughout the melting cycle. In addition a temperature gradient forms within the scrap. Temperatures must be controlled to prevent damage to the scrap bucket and in order to prevent burning or sticking of fine scrap within the bucket. Scrap temperatures can reach 315-450℃ (600-850°F) though this will only occur at the hot end where the offgas first enters the preheater. Savings are typically only in the neighborhood of 18-23kW · h/t. In addition, as operations become more efficient and tap-to-tap times are decreased, scrap preheating operations become more and more difficult to maintain. Scrap handling operations can actually lead to reduced productivity and increased maintenance costs. At Badische Stalwerke the energy savings due to scrap preheating were decreased by 50% when tap-to-tap time was reduced by one third.

Some of the benefits attributed to scrap preheating are increased productivity by 10%-20%, reduced electrical consumption, removal of moisture from the scrap, and reduced electrode and refractory consumption per unit production. Some drawbacks to scrap preheating are that volatiles are removed from the scrap, creating odors and necessitating a post-combustion chamber downstream. In addition spray quenching following post-combustion is required to prevent recombination of dioxins and furans. Depending on the preheat temperature, buckets may have to be refractory lined.

5.5.2 Preheating with Offgas

Preheating with offgas from the EAF requires that the offgas be rerouted to preheat chambers which contain loaded scrap buckets. The hot gases are passed through the buckets thus preheating the scrap. For tap-to-tap times less than 70 minutes the logistics of scrap preheating lead to minimal energy savings that will not justify the capital expense of a preheating system. Typical savings are in the range of 15-20kW · h/t. Some examples of preheating systems are given in Fig. 5.20 and Fig. 5.21.

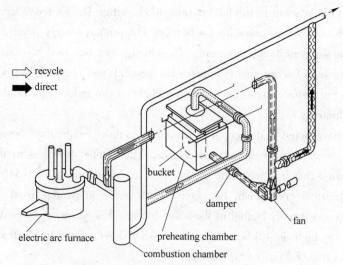

Fig. 5.20 NKK scrap preheater (Courtesy of NKK)

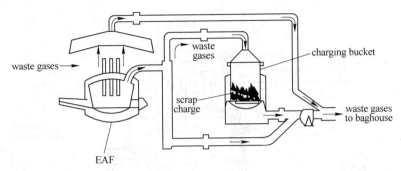

Fig. 5.21 Scrap preheating in the scrap bucket (Courtesy of Center for Materials Production)

5.5.3 Natural Gas Scrap Preheating

Natural gas scrap preheating originated in the 1960s and usually involves a burner mounted in a refractory lined roof which sits over the top of the scrap bucket. Scrap is typically preheated to 540-650℃. Above 650℃, scrap oxidation becomes a problem and yield loss becomes a factor. Advantages of this form of scrap preheating are that the preheating process is decoupled from EAF operations and as a result the process is unaffected by tap-to-tap time. However, heat is provided by natural gas as opposed to offgas and as a result an additional cost is incurred.

One of the primary concerns with scrap preheating is that oil and other organic materials associated with the scrap tend to evaporate off during preheating. This can lead to discharge of hydro carbons to the atmosphere and foul odors in the shop environment. In some Japanese operations, this has been remedied by installing a post-combustion chamber following scrap preheating operations. In one such operation scrap preheating is conducted in conjunction with a furnace enclosure. It is reported that power savings are 36-40kW · h/t and electrode consumption is reduced by 0.4-0.6kg/t. It is not specified what the additional capital costs and maintenance costs were for this system. In North America, most operations find that the savings offered by conventional scrap preheating do not compensate for the additional handling operations and additional maintenance requirements.

5.5.4 K-ES

The K-ES process is a technology developed jointly by Klockner Technology Group and Tokyo Steel Manufacturing group. Subsequently the rights to the process were acquired by VAI. The process uses pulverized or lumpy coal in the bath as a source of primary energy. Oxygen is injected into the bath to combust the coal to CO gas. The CO gas is post-combusted in the furnace freeboard with additional oxygen to produce CO_2. Thus a large portion of the calorific heat in the process is recovered and is transferred to the bath. In addition the stirring action caused by bottom gas injection results in better bath mixing and as a result accelerated melting of the scrap. The process is shown schematically in Fig. 5.22. Fig. 5.23 shows the carbon and oxygen injectors used in K-ES. Fig. 5.24 presents oxygen consumption versus electrical power consumption.

162   5 Electric Furnace Steelmaking

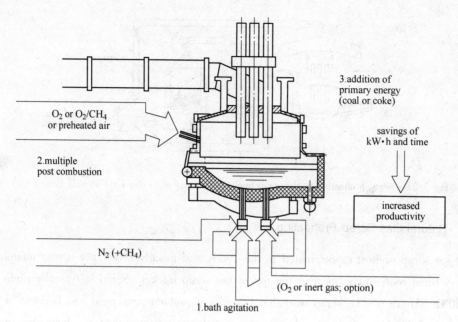

Fig. 5.22 K-ES process (Courtesy of Voest Alpine Ind.)

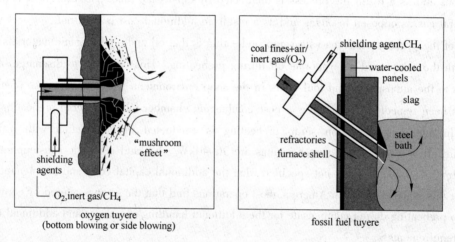

Fig. 5.23 K-ES carbon and oxygen injectors (Courtesy of Voest Alpine Ind.)

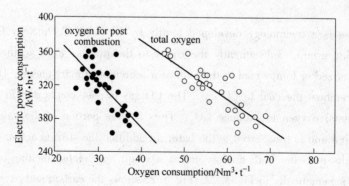

Fig. 5.24 Total oxygen consumption versus electrical power consumption for K-ES operations

5.5.4.1 Historical Development

The first K-ES installation was at Tokyo Steel in 1986, where a 30 tons EAF was converted to run trials on the process. A productivity increase of 20% and an electrical savings of approximately 110kW·h/t resulted. Thus the process concept was proven on an industrial scale.

In 1988, Ferriere Nord in Italy decided to install the K-ES process on its 88 tons EAF in Osoppo. At that time Ferriere Nord was producing approximately 550,000 tons per year of steel on a 1975 vintage EAF that was originally designed to produce 220,000 tons per year. This had been accomplished through a combination of long arc/foamy slag practice, high oxygen utilization ($32Nm^3/t$), water-cooled furnace roof and walls and oxy-fuel burners. The first K-ES heat took place early in 1989 following the installation of post-combustion lances, coal injection equipment and bottom injection tuyeres. This process continues to operate at Ferriere Nord.

In December 1989 another K-ES facility was installed on a 82 tons oval tapping furnace at Acciaierie Venete in Padua, Italy. This operation differs in that the A phase electrode is hollow and is used for carbon injection. This installation has seven post-combustion lances that are mounted into the furnace wall, through the water-cooled panels. Five bottom tuyeres are located to form a pitch circle positioned between the furnace wall and the electrode pitch circle. The operation at Venete has achieved a productivity of 2.2t/MW. Tap-to-tap times average 54 minutes even though the transformer is only capable of supply a maximum of 45 MW.

5.5.4.2 Results

At Tokyo Steel the average power consumption for a set of trials was 255kW·h/t with an average coal injection rate of 24kg/t. This gave an electrical power replacement of 4-5.5kW·h/kg of coal injected. In addition the melting time was reduced from 60 minutes to 48-50 minutes. Ferrosilicon and ferromanganese consumption was also decreased by 1.1kg/t. Results are presented in Table 5.6.

Table 5.6 K-ES analysis based on Tokyo Steel results

Material	Consumption/t Liquid steel	Difference
Power on time	58 minutes	−26 minutes
Tap-to-tap time	79 minutes	−26 minutes
Production	476,000t/year	97,000t/year
Coal	32kg/t	+27kg/t
Oxygen	$63Nm^3/t$	$+53Nm^3/t$
Electricity	330kW·h/t	−150kW·h/t
Lime		+5.0kg/t
Electrodes		−0.3kg/t
FeMn		−1.0kg/t
Inert gas	$6.0Nm^3/t$	$+6.0Nm^3/t$
Production increase		+33%

Following installation and testing of the K-ES system at Tokyo Steel, an 88 tons furnace at Ferriere Nord was converted to K-ES operation. Lumpy coal is added in the charge and additional powdered coal is injected into the bath. Oxygen consumption is $50 Nm^3/t$. The distribution of this oxygen is as follows: $13 Nm^3/t$ is injected to the bath via tuyeres, $13 Nm^3/t$ is injected to the bath via an oxygen lance and $18 Nm^3/t$ is injected into the freeboard through post-combustion lances. Typically carbon consumption is on the order of 20kg/t unless the charge is 30% hot pig metal, in which case the carbon consumption is 13kg/t. Electrical energy consumption decreased by 60kW·h/t with the K-ES operation. An energy balance showed that the electrical savings were highly dependent on efficient post-combustion of the process gases in the freeboard.

Best results at Ferriere Nord were achieved at conditions of 180kW·h/t, $53 Nm^3 O_2/t$ (1850 scf O_2/t), 30% pig iron and tap-to-tap time of 35 minutes and at conditions of 255kW·h/t, $40 Nm^3 O_2/t$ (1400 scf O_2/t), 100% scrap and tap-to-tap time of 53 minutes.

Similar results have been obtained at Acciaerie Venete using a Fuchs OBT furnace with K-ES.

5.5.5 Danarc Process

The Danieli Danarc process combines high impedance technology with high chemical energy input to the furnace in order to achieve high productivity and energy efficiency. The first installation of this technology was at Ferriere Nord in Italy. Features for chemical energy input are very similar to K-ES. Bottom tuyeres are used to inject oxygen and carbon. Sidewall lances are also used. Post-combustion oxygen is supplied via burners. Fig. 5.25 shows a top view of the Danarc furnace at Ferriere Nord.

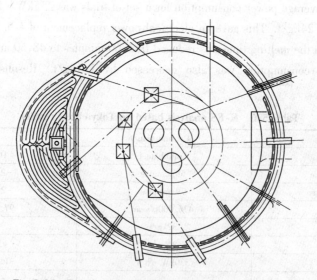

Fig. 5.25 Top view of a Danarc furnace (Courtesy of Danieli)

The purpose of tuyeres installed in the furnace bottom is to distribute the oxygen throughout the furnace in order to maximize the decarburization rate. With just oxygen lancing, the area around the lance becomes depleted of carbon and some of the oxygen reacts with iron. CO is generated mostly in one part of the furnace around the injection point. With multiple tuyeres, CO generation is spread out within the furnace and the potential for heat recovery via post-combustion is greater. Injection at several points also gives good bath mixing, an added benefit. Multiple sidewall carbon injectors allow for good control of slag foaming across the whole surface of the bath.

Natural gas, nitrogen and carbon dioxide are used as shroud gas for the oxygen tuyeres. Wear rates are approximately 0.5mm per heat.

Traditionally, a series reactor has been installed on the primary side of the furnace transformer to allow for high impedance operation. This allows the furnace to operate at long arc (i.e. high voltage) and low electrode current which gives high arc stability, improved power input to the bath and reduced electrode consumption.

Alternatively, the saturable reactor provides a method to reduce both current and reactive power fluctuation. This reduces the electrodynamic stresses on the furnace transformer secondary circuit and reduces the flicker level in the supply network. The main purpose of the saturable reactor is to control the reactance so as to minimize current fluctuations. The saturable reactor acts as a variable reactance controlled by the excitation current. The current control performance becomes similar to a DC furnace with thyristor controlled current. Several advantages resulting from the use of the saturable reactor are that current fluctuations and electrodynamic stresses are limited, reactive power fluctuations are reduced resulting in less flicker, energy transfer to the melt is improved, reduced electrode consumption, and tap changing is not necessarily required.

The following results are reported for Ferriere Nord for a charge makeup of 82% scrap and 18% cold pig iron. Tap-to-tap time of 50 minutes was achieved, with an aggregate power-on time of 40 minutes. Power consumption was 270kW · h/t liquid. Electrode consumption was 1.6kg/t liquid. Total oxygen consumption was 50Nm^3/t liquid while total natural gas consumption was 10Nm^3/t liquid. Total carbon introduced was 10kg/t liquid. The resultant tap temperature was 1640℃. For 100% scrap operation, the power consumption increases by 23kW · h/t liquid.

Trials have also been run using hot metal, with 70% scrap and 30% hot pig iron as part of the charge. The results were a tap-to-tap time of 45 minutes and a power-on time of 38 minutes. Power consumption was 160kW · h/t liquid. Electrode consumption was 1.0kg/t liquid. Total oxygen consumption was 40Nm^3/t liquid and total natural gas consumption was 2.2Nm^3/t liquid. Air supplied to burners was 9Nm^3/t liquid and total carbon injected was 14.9kg/t liquid. Resultant tap temperature was 1680℃.

5.5.6 Fuchs Shaft Furnace

The need to reduce the amount of power input into EAF operations lead to a prototype study of a shaft furnace at Danish Steel Works Ltd (DDS). The concept was to load scrap into a shaft where it would be preheated by offgases exiting the EAF. The scrap sat in a column at one end of the furnace and was constantly fed into the furnace as the scrap at the bottom of the column melted away.

In January 1990, a production shaft preheater was retrofitted to one of DDS's two 125 tons EAFs. DDS stopped using the shaft after less than two years of operation. This was partly due to the problems of maintaining a suitable scrap flow in the shaft. In addition, Danish tariff regulations on electricity made a three shift per day operation uneconomical. As a result DDS could not move to a single furnace operation which would make full use of the preheating shaft.

5.5.6.1 Single Shaft Furnace

The Fuchs shaft furnace concept was installed at Co-Steel Sheerness in England in 1992. This installation is the outcome of work done in Denmark at DET Danske Stalvalsevaerk on first and second versions of the process. The process has a shaft with a conventional oval EAF bottom with three phase AC arcs. Fig. 5.26 illustrates the shaft. The relatively short shaft is scrap bucket fed. Unlike the EOF, there are no movable fingers and the scrap descends continuously into the bath where the arcs and oxygen produce the unrefined metal. The Fuchs shaft preheater at Sheerness consists of a reverse taper (larger at bottom) shaft which sits on the furnace roof, offset from the furnace centerline.

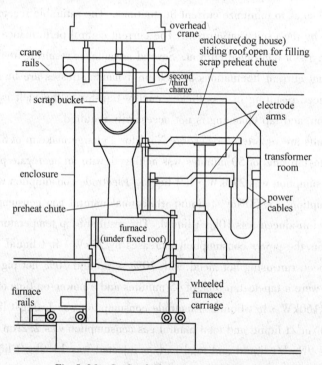

Fig. 5.26 Co-Steel Sheer-ness shaft furnace

The furnace shell is mounted on a frame and is tilted by means of four hydraulic cylinders mounted in each corner of the frame, enabling the furnace shell to be lowered by one foot to the first bucket charging position or to be tilted into the tapping or deslagging positions. The furnace runs on four tracks, two in the center and two on the outside. The furnace water-cooled roof, the shaft and the electrode gantry are fixed and do not tilt.

Energy input to the furnace is from an 80MV · A transformer, 6 MW oxy-fuel burners and a watercooled oxygen lance.

A typical operating cycle is as follows. The heat cycle is started by charging the first basket of scrap. The furnace is lowered one foot onto a bumper and the furnace car is moved west to the charging position where the first bucket is held on a bale arm support structure. The bucket is opened remotely via claw-type actuators. The first charge is approximately 44 tons of scrap.

Once the furnace has been charged it is moved back under the roof and is raised back to the upper position. The second bucket is then charged to the shaft. The second charge is approximately 35 tons. The complete charge cycle for the first two buckets of scrap is typically less than 2 minutes. Meltdown commences and after approximately four minutes the electrodes are raised and the third and final scrap bucket is charged to the shaft. Meltdown then proceeds uninterrupted and the furnace taps 99 tons at 1640℃ with an average power-on time of 34 minutes.

The Fuchs shaft has claimed benefits ranging from reduced EAF dust (due to dust sticking to the scrap in the shaft) to reduced exhaust fan requirements due to lower gas volumes. As all of the scrap is melting in, the zinc quickly volatizes, resulting in furnace dust that is high in zinc content. For a 120 tons furnace with an 80MV·A transformer, Fuchs predicts a tap-to-tap time of 56 minutes with an electrical power consumption of approximately 310kW·h/t. The projections anticipate that the equivalent of 72kW·h/t of energy will be recovered from the offgas.

The results reported by Co-Steel Sheerness PLC have indicated the following:

1. Liquid steel yield, due to high FeO reversion from the slag, resulted in an average of 93.5%.

2. The volume of flue dust produced by the shaft furnace was 20% lower than for the conventional furnace (14kg/t billet compared to 18kg/t billet).

3. The flue dust chemistry was changed with shaft furnace operation. The zinc oxide content rose from 22% to 30%. In addition the lime content was decreased from 13% to 5%.

4. Fan power requirements for the fume extraction fans were decreased from 19.3kW·h/t billet to 10kW·h/t billet due to lower gas volumes.

Table 5.7 presents results for various trials conducted at Co-Steel Sheerness. As can be seen, the amount of burner power was increased considerably in the most recent set of trials. The burner efficiency that was achieved was much higher than that in a conventional furnace as the flame contacted cold scrap for a longer duration. In addition the CO content in the offgas was monitored in order to adjust the oxygen flow to the burners to promote CO post-combustion at the base of the shaft.

Table 5.7 Results of shaft furnace operating trials at Co-Steel sheerness

Items	Old furnace	Standard shaft charge	Trial shaft charge
Electrical energy /kW·h·t^{-1} billet	467	327	250
Burner oxygen /Nm^3·t^{-1} billet	3.0	20.0	3.2
Lance oxygen /Nm^3·t^{-1} billet	10.9	10.0	5.0
Burner natural gas /Nm^3·t^{-1} billet	1.5	10.0	16.0

Fuchs has other single shaft furnaces, of similar design, operating in Turkey, China, Malaysia and at North Star Steel in Arizona. The latter, Fig. 5.27, is powered with a DC transformer and is equipped with an ABB DC bottom.

Fig. 5.27   Shaft furnace in operation at North Star Steel, Kingman, Arizona

5.5.6.2   The Double Shaft Furnace (DSF)

In order to increase the production capacity of a furnace with one transformer to over 1,000,000 tons per year, the double shaft furnace was developed. There are currently two in operation in Europe, SAM in France and ARBED in Luxembourg (Fig. 5.28). Both are 95 tons AC high impedance furnaces which, when fully utilized, deliver a liquid steel capacity of over 1,000,000 tons per year. The double shaft furnace utilizes one transformer and one set of electrodes for both furnace shells.

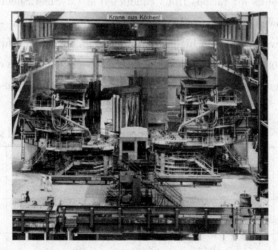

Fig. 5.28   Double shaft furnace in operation at ARBED (Courtesy of Fuchs)

North Star BHP Steel Ltd. started up the first double shaft furnace in the United States. This furnace, with a tap weight of 180 tons, has a rated capacity of 1,700,000 tons per year. The AC transformer is rated at 140MV · A and will operate with a secondary voltage of 1200-1300 V. All (100%) of the scrap charged can be preheated. A savings in electric energy consumption of between 100-120kW · h/t is projected in comparison with a single shell EAF based on 100% scrap charge. North Star BHP Steel plans to utilize DRI/HBI and/or pig iron as part of the charge. This type furnace would also be well suited for the use of iron carbide.

The trend toward the use of more fossil energy by electric arc furnace steelmakers will result in more latent energy in the offgas, primarily in the form of CO. Measurements indicate a value of energy in the offgas of 170-190kW · h/t in a standard single shell EAF operation. In addition, instead of using energy to circulate water in the duct system to cool the hot offgases, this energy could instead be utilized to heat and melt the scrap, as is done in shaft furnaces.

The double shaft furnace at SAM, Table 5.8, currently operates with a typical scrap charge that is 25% heavy scrap, 30% shredded, 5% municipal recycled, 15% galvanized sheet, 10% turnings and 15% light scrap. Three charge buckets are used. A round bucket holding 50% of the charge is used to charge the furnace. Two rectangular buckets each containing 25% of the charge are used to charge the shaft. While one shell (A) is tapping, the electrodes are moved to the other shell (B) to begin meltdown. Bore-in proceeds at a voltage of 750V. Meltdown then proceeds at 900V. Six burners located at the bottom of the shaft aid meltdown. Vessel A, once finished tapping, is charged with 75% of the total charge. The burners in this vessel are placed on high fire to help preheat the scrap. Vessel B completes meltdown and its offgases are directed to the vessel A to aid in scrap preheating. The remaining 25% of the scrap is now charged to the shaft. Once vessel B is ready to tap, the electrodes are moved back over to vessel A to start another meltdown cycle.

Table 5.8 Furnace data of the AC double shaft furnace at SAM

Furnace capacity	115t
Furnace diameter	6.3m
Tapping weight	99t
Liquid heel	11t
Transformer capacity	95MV · A
Active power	60MW
Electrode diameter	600mm
Oxy-gas burners	7×3MW (ea)
Oxy-carbon lances	2×35Nm^3/min
Bottom stirring	5 elements-nitrogen

Consumption figures reported by SAM and ARBED are presented in Table 5.9 and Table 5.10, respectively.

Table 5.9 Reported consumption figures at SAM

Items	Best day	Best 4 days
Charge mix	Scrap 100%	Scrap 100%
Electrical energy	320kW·h/t	338kW·h/t
Electrode consumption	1.3kg/t	1.45kg/t
Oxygen lance	10.3Nm^3/t	11.8Nm^3/t
Oxygen burners	11.8m^3/t	12.7Nm^3/t
Gas	6.5Nm^3/t	6.2Nm^3/t
Carbon charge	7.5kg/t	7.6kg/t
Carbon foamy slag	6.7kg/t	5.4kg/t
Lime	35.2kg/t	39.5kg/t
Power on time	40min	45.5min
Liquid yield	91.5%	

Table 5.10 Reported consumption figures at ARES Schifflange

Electrical energy	298kW·h/t
Electrode consumption	1.23kg/t
Oxygen	22.0Nm^3/t
Gas	5.9Nm^3/t
Power on time	39.2min

5.5.6.3 The Finger Shaft Furnace

Various studies have shown that the greatest proportion of the energy leaving the furnace in the offgas occurs during flat bath operations. This is due to the absence of scrap to absorb energy from the offgas as it exits the furnace. A large potential for energy recovery exists if the offgas can be used to preheat scrap. For operations with continuous feed of high carbon iron units (e.g. DRI or iron carbide), the amount of energy escaping in the offgas will be higher yet due to the higher levels of carbon monoxide in the offgas. Finger shaft furnace installations with tap weights ranging from 77-165 tons are located at Hylsa, Cockerill Sambre, Swiss Steel AG and Birmingham Steel.

To use the energy of the offgas efficiently and to reduce the electrical power requirements needed to obtain short tap-to-tap times, Fuchs developed the finger shaft furnace at Hylsa in Monterrey, Mexico. This furnace, tapping 150 tons, maintains a 44 tons liquid heel and is powered with a 156MV·A DC transformer utilizing the Nippon Steel water-cooled DC bottom electrode (3 billet bottom electrode). Reported consumption figures are presented in Table 5.11. The furnace charge is 45% scrap and 55% DRI charged during the 50 minutes power-on time of the furnace. Half of the scrap is charged onto a water-cooled finger system in the shaft during the refining period of the previous heat. The other half of the scrap is charged onto the fingers after the first charge has been dropped into the furnace and power is applied.

5.5 New Scrap Melting Processes

Table 5.11 Reported consumption figures at Hylsa

Charge mix	50% scrap, 50% DRI
Electrical energy	394 kW·h/t
Electrode consumption	0.9 kg/t
Oxygen lance	20 Nm^3/t
Oxygen burners	8.1 Nm^3/t
Natural gas	3.6 Nm^3/t
Carbon charge	9 kg/t
Carbon foamy slag	4.5 kg/t
Power on time	48 min
Tap temperature	1620 ℃

The furnace roof and shaft sit on a trolley which moves on rails mounted on the slag door and tapping sides of the furnace. To initiate a heat, the electrode is raised and rotated away from the furnace. The shaft is moved into position over the furnace center and the fingers are retracted so that the preheated scrap is charged onto the hot heel. The roof is then moved back and the electrode is returned to its operating position. A second scrap charge is added to the shaft and DRI is fed continuously throughout meltdown. Once the first scrap charge is melted, the second charge is added to the center of the furnace. Initial bore-in utilizes a voltage of 400V and is increased to 550V once the electrode has penetrated to about one metre. During meltdown a voltage of 635V is used. During meltdown of the second scrap charge and refining, the first charge for the next heat is preheated in the shaft. The DRI, containing 2.3% carbon, is cold and continuously charged. Oxy-fuel burners are used during scrap meltdown. Future plans call for running the burners with excess oxygen for post-combustion of CO.

A finger shaft furnace is now operating at Swiss Steel AG (Von Roll). This furnace was converted in only five weeks. It is powered with a high impedance AC transformer. They have seen a 55% improvement in productivity and are tapping 79 tons of steel every 37 minutes. Reported consumption figures are presented in Table 5.12.

Table 5.12 Reported consumption figures at Swiss Steel AG

Charge mix	100% scrap
Electrical energy	260 kW·h/t
Electrode consumption	1.3 kg/t
Oxygen lance	14.4 Nm^3/t
Oxygen burners	11.8 Nm^3/t
Natural gas	5.1 Nm^3/t
Carbon charge	5.5 kg/t
Carbon foamy slag	2.7 kg/t
Power on time	28 min
Tap temperature	1620 ℃

This operating data indicates that 30 minutes tap-to-tap times are likely to be achieved in the future.

Paul Wurth S. A. has installed a 154 tons DC finger shaft furnace at Cockerill Sambre in Belgium under license from Fuchs. The unique feature of this finger shaft furnace is that 20%-50% of the charge can be liquid iron with a carbon content of 4%. The operation utilizes the Paul Wurth proprietary hot metal charging system.

This operation utilizes four water-cooled billets as the bottom electrode. The hot metal charging system is a technology developed by Paul Wurth based on trials conducted at various ISCOR facilities in South Africa. Hot metal is charged through a side launder in the furnace as shown in Fig. 5.29. This system allows for continuous charging of 44-60 tons of hot metal to the furnace over a period of 15-20 minutes. The furnace is located within an enclosure to minimize fume emissions to the shop. Fig. 5.30 shows the hot metal ladle charging system operating sequence. Reported consumption figures at Cockerill Sambre are reported in Table 5.13.

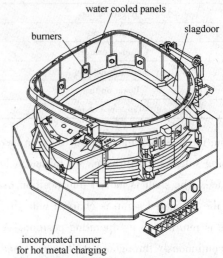

Fig. 5.29 Shaft furnace bottom with runner for hot metal charging (Courtesy of Paul Wurth, S. A.)

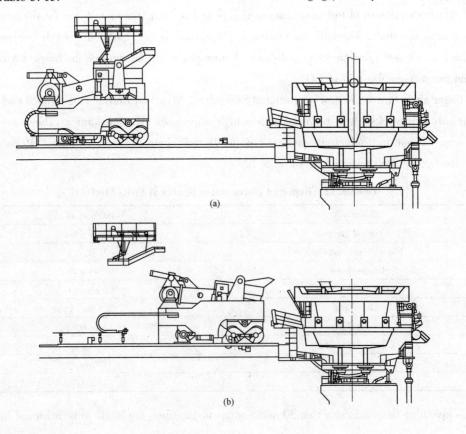

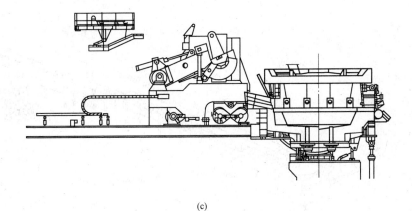

(c)

Fig. 5.30 Operating sequence to charge hot metal
(a) Ladle with cover in place; (b) Cover removed and ladle to furnace;
(c) Charging hot metal (Courtesy of Paul Wurth, S. A.)

Table 5.13 Reported consumption figures at Cockerill Sambre

Items	34% hot metal charge	100% scrap charge
Electrical energy	187kW · h/t	290-310kW · h/t
Power on to tap	<40min	
Oxygen		23-27Nm^3/t
Natural gas		5.5-7.3Nm^3/t
Carbon charge		9-13.5kg/t
Power on time		50-55min
Tap temperature		1620℃

A replacement equation for the equivalent value of hot metal in the furnace has been developed by Paul Wurth based on various plant trials and is presented below.

1t hot metal 122.7kg lime = 0.92t scrap 145kg coal 1300kW · h electricity

5.5.7 Consteel Process

The Consteel process was developed by Intersteel Technology Inc. which is located in Charlotte, North Carolina. This is another process that is based on recovering heat from the offgas. In this case, the scrap enters a long preheater tunnel and is preheated by the offgas as it travels to the furnace. The scrap is moved through the tunnel on a conveyor and is fed continuously to the EAF. The offgas flows counter-current to the scrap. The EAF maintains a liquid heel following tapping. Fig. 5.31 illustrates the key system components.

5.5.7.1 Historical Development

The Nucor Steel Corporation was the first company to commit to an industrial application of the Consteel technology. In 1985 Consteel was retrofitted into an existing Nucor plant located in Dar-

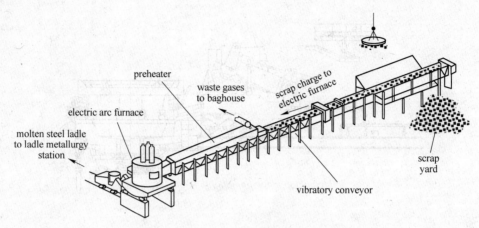

Fig. 5.31　Key components of the Consteel process

lington, South Carolina. Due to space restrictions the installation was designed with a 90° bend in the scrap conveyor and a relatively short preheater section. This was reported to result in scrap feeding and preheating problems. The unit was operated for eighteen months and was then shut down.

Several key concepts were confirmed by this prototype which was put into operation in 1987. A consistent heel of hot metal acted as a thermal flywheel, increasing efficiencies of scrap melting. It was demonstrated that keeping the bath temperature within a proper range ensured a constant equilibrium between metal and slag and a continuous carbon boil, which resulted in a bath which was homogeneous in temperature and composition. The foaming of the slag could be continuously and precisely controlled and was very important for successful operation.

The first greenfield demonstration of the Consteel process was at AmeriSteel in Charlotte, North Carolina. This layout included a new meltshop that was built parallel to an existing one. In this case the continuous scrap feed system and the scrap preheater were all in line. The scrap was taken from railcars to a loading station through a preheater to a connecting car and then into the furnace. The preheater design included an afterburner for control of carbon monoxide emissions though this feature was never put into operation. The scrap preheater was designed to heat scrap up to 700℃. Particular emphasis was placed on providing a tight seal on the scrap preheater. This was accomplished by providing a water seal in the preheater. The key characteristics of this facility have been described.

The Contifeeding system consists of three conveyors in cascade, each 1.5 metres (5ft) ×0.3 metres (1ft) deep×60 metres (200ft) long. A refractory lined tunnel covers the conveyor and a water seal prevents outside air from leaking between the cover and the conveyor pans. Shredder scrap, #1 scrap, turnings and light structural scrap are charged to the furnace on the conveyor. A leveller bar at the scrap charge end maintains a maximum scrap height of 0.45 metres (18in) on the conveyor.

The scrap preheater is 24 metres (80ft) in length with 60 natural gas burners of 7.0Nm^3/min (250 scfm) capacity mounted in the preheater roof. Preheat temperatures up to 700℃ are achieved using the furnace offgas and the burners. The burners were later removed as sufficient preheating was achieved using only furnace offgas.

The furnace proper is a 75 tons EBT EAF designed to tap 40 tons and retain a 30-35 tons heel. The scrap feed rate is approximately 680kg/min (1500lbs/min). For cold startup the furnace is top charged and this charge is melted in to provide the liquid heel; a slag door burner is used to accelerate meltdown for these heats. Electrical energy is supplied by a 30MV · A transformer through 0.5m (20in) electrodes, with resultant tap-to-tap times of approximately 45 minutes. Other facilities include carbon and oxygen bath injection for producing a foamy slag and a pneumatic lime injection system.

In addition to the operation at AmeriSteel, several other plants have installed Consteel at their facilities including Kyoei Steel (Nagoya), Nucor Steel Darlington and New Jersey Steel.

5.5.7.2 Key Operating Considerations

The key to obtaining good operating results with Consteel is based on controlling several operating parameters simultaneously. These are bath temperature, scrap feed rate and scrap composition, oxygen injection rate, bath carbon levels, and slag composition.

It can be seen that any type of imbalance in any of these operating parameters will have a ripple effect through the whole process. Generally, the hot heel size is 1.4 tons times the power input in MW. Some operations use process models to track the process inputs. Slag composition is very important because without good slag foaming it will be difficult to bury the arc. This will result in large heat losses and potentially furnace damage. Continuous evolution of CO from the bath is crucial to maintaining scrap preheat temperatures. Approximately 70%-75% of the CO generated in the furnace is available as fuel in the preheater. A reducing atmosphere is maintained in the first 30% of the preheater in order to control scrap oxidation. Complete combustion of CO and VOCs evolving from the scrap is achieved prior to the fluegas exiting the preheater. The FeO content of the slag is maintained around 15% which is suitable for good slag foaming.

A variety of scrap types can be fed to the Consteel operation. The primary stipulation is that the size is less than the conveyor width in order to minimize scrap bridging in the feed system. Bundles can be used but will not benefit from preheating. Best results are obtained when using loose, shredded scrap which is light and has a large surface area for heat transfer from the hot offgases. High carbon scrap such as pig iron can be used but must be distributed within the charge in order to maintain a fairly uniform carbon level in the bath. HBI can be charged to the preheater along with scrap. DRI should only be charged into the section of the preheater with a reducing atmosphere. The key to good operation in all cases is to maintain homogeneity of the scrap feed so that bath composition remains within preset limits. Scrap density must also be optimized so that the desired residence time in the preheater shaft can be attained thus ensuring sufficient preheating of the scrap. Charge permeability also affects the amount of preheat achieved.

The following summary, Table 5.14, from Intersteel lists the latest data on existing and future installations.

Table 5.14 Consteel operating results

Facility	AmeriSteel Charlotte	Kyoei Nagoya	Nucor Darlington	New Jersey Steel	ORI Martin Italy	N.S.M. Thailand
Start-up date	Dec. 1989	Oct. 1992	Sept. 1993	May 1994	On Hold	End 1997
Productivity	54t/h	140t/h	110t/h	95t/h	87t/h	229t/h
EAF type	AC	DC	DC	AC	AC	AC
New/Retrofit	New	New	New	Retrofit	Retrofit	New
Transformer power	24MW	47MW	42MW	40MW	31MW	95MW
Power	370kW·h/t	300kW·h/t	325kW·h/t	390kW·h/t		
Oxygen	22Nm^3/t	39Nm^3/t	33Nm^3/t	23Nm^3/t		
Electrodes	1.75kg/t	1.14kg/t	1.0kg/t	1.85kg/t		
Yield	93.3	94	93	90		

Note: some of the above figures represent unit consumption records.

5.5.8 Twin Shell Electric Arc Furnace

5.5.8.1 Historical Development

One of the new technologies that seems to be generating much interest is the twin shell furnace. Essentially this technology is similar to conventional scrap preheating with the exception that scrap preheating actually takes place in a furnace shell as opposed to a scrap bucket.

The original work done on in-furnace scrap preheating was in Sweden, where SKF operated an installation which had a single power supply and two furnace shells. In the early 1980s Nippon Steel developed a twin shell process for the production of stainless steel (Fig. 5.32). At the current time, several furnace manufacturers are producing twin shell furnaces. The key goals of a twin shell operation are equal to those of other developing technologies, but in addition the cycle times are similar to those for BOF operations due to the minimization of power-off times.

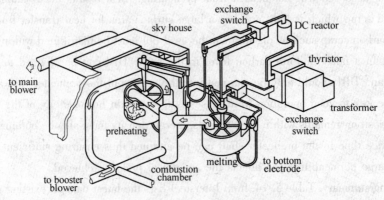

Fig. 5.32 Schematic of the DC twin shell EAF at Nippon Steel

5.5.8.2 Process Description

This type of operation consists of two furnace shells and one set of electrode arms that are used alternately on one shell and then the other. The scrap is charged to the shell that is not melting and the scrap is preheated by the offgas from the melting furnace. Supplemental burner heat can also be used. The result is that the scrap is preheated in the furnace prior to melting. The more the scrap is preheated, the greater the energy savings.

Some early designs proposed a gantry that moved along a rail from one shell to the other. Most designs now use a furnace roof and electrode arms that can swing between two positions for the shells. Generally a twin shell installation will consist of two identical vessels with a lower shell, an upper shell and a roof and one set of electrode arms and lifting supports with one conventional power supply.

It is interesting to note that several of the recent orders for twin shell installations have opted for DC operation. This obviously has some inherent benefits as there is only one electrode arm to rotate between the two operating positions. Another variation introduced is that some operations have an electrode mast and arm for each furnace and have swithgear allowing the sharing of a common transformer.

There have been several operating modes suggested for the twin shell operation. The Nippon Steel operating cycle consists generally of two phases; scrap preheating and melting. The furnace is only charged one time and thus the heat size is smaller than the actual capacity of the furnace. In the case of the Nippon Steel installation at POSCO, the operation is a two charge operation where 60% of the total charge is preheating during the melting phase on the second furnace. Thus the second charge (40% of the total) is not preheated in the furnace.

The operating cycle proposed by Clecim is similar to that at POSCO. The Mannesmann Demag operating cycle is quite different in that the operation uses two charges but the power is alternated between furnaces between charges. Thus one furnace melts its first charge while the other preheats its first charge. Once the first charge is melted, the second charge is dropped and the power is diverted to the first preheated charge. Thus the second charge is preheated as well as the first during the preheating cycle for the second charge; oxygen lancing or oxy-fuel burners can also be used to accelerate preheating and melting. This operating method maximizes the recovery of heat from the offgas but requires very precise process control to keep from freezing the molten heel in the furnace during the preheat phase.

5.5.8.3 Operating Results

A Nippon Steel

In this operation, there is only one charge per heat. While one charge is being melted, a second charge is being heated in the second shell. Combustion gas from a burner on the second shell is mixed with offgas from the first shell to give a constant temperature of 899℃ (1650℉) for preheating in the second vessel. The remaining offgas from both shells is used to preheat scrap in a scrap bucket. The net electrical power input requirement is reported at 260kW・h/t which is 29% lower than for the two conventional furnaces that were replaced by the twin shell operation.

B Davy-Clecim

The Clecim installation at Unimetal Gandrange tapped its first heat in July 1994. This twin shell operation is DC with the Clecim water-cooled billet bottom electrode. The shells have a nominal capacity of 165 tons with a nominal diameter of 7.3 metres. Each shell has a working volume of approximately 200m^3 which enables the furnace to operate with a single charge each heat. The rated installed transformer capacity is 150MV・A. The maximum secondary voltage is 850V. Maximum power input is 110MW. Each shell has four bottom billet electrodes. The graphite electrode swings between the two furnaces as required. Each shell has a manipulator with consumable oxygen and carbon injection lances. Initially, integrated steelmaking operations continued to operate which led to large furnace delays due to the logistics of movements within the shop. This operation experienced problems with arc flare due to the configuration of the anode bus tubes. The configuration was altered and this problem was eliminated.

The ProfilArbed installation started up at the end of 1994 and is shown in Fig. 5.33.

Fig. 5.33 Twin shell furnace operation at ProfilArbed

Several recent North American twin shell installations include Steel Dynamics, Gallatin Steel, Tuscaloosa Steel and Nucor-Berkeley. Fig. 5.34 shows the operation at Tuscaloosa Steel.

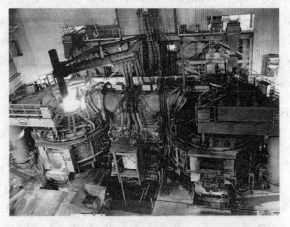

Fig. 5.34 Twin shell furnace operation at Tuscaloosa Steel

5.5.9 Processes under Development

Based on the success of several of the scrap preheat technologies (Consteel, Fuchs Shaft Furnace, EOF), several new shaft type technologies are currently in the development stage. These are outlined in this section.

5.5.9.1 Ishikawajima-Harima Heavy Industries (IHI)

IHI is currently developing a shaft type preheat furnace based on twin electrode DC technology. The first commercial installation has started up at the Utsunomiya plant of Tokyo Steel. The DC furnace itself is oval in shape with two graphite electrodes and two bottom electrodes consisting of conductive hearth brick (as per ABB's DC furnace design). There are two DC power supplies which are individually controlled. The power feeding bus is arranged so that the two arcs will deflect towards the center of the furnace, thus the energy of the arcs will be concentrated at the center of the furnace and the thermal load to the furnace walls will be low compared to a conventional furnace. As a result refractory walls are used instead of water-cooled panels thus reducing heat loss. Scrap is charged to the furnace between the electrodes. The furnace will maintain a large hot heel (110 tons heel, 140 tons tap weight) so that uniform operating conditions can be maintained (similar to the Consteel concept). Steel is tapped out periodically via a bottom taphole in the furnace. A diagram of the furnace is given in Fig. 5.35.

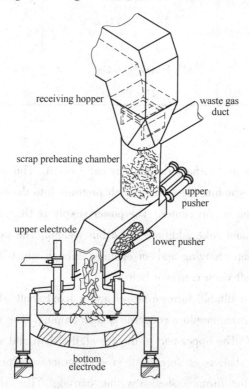

Fig. 5.35 IHI shaft furnace

The scrap charging system consists of two main components, the preheat chamber and the charging equipment. The scrap is fed into the upper part of the chamber from a receiving hopper. The exhaust gas from the furnace flows up through the chamber, preheating the scrap. In the pilot plant, scrap preheat temperatures as high as 800℃ were achieved. Gas exit temperatures from the chamber were as low as 200℃. At the base of the preheat chamber are two pushers. These operate in two stages, allowing scrap to feed into the furnace at a constant rate. Offgas leaves the top of the preheat chamber and flows to a bag filter. Some gas can be recycled to the furnace to regulate the inlet gas temperature to the preheater.

Scrap is fed continuously to the furnace until the desired bath weight is achieved. This is followed by a short refining or heating period leading to tapping of the heat. Power input is expected to be almost uniform throughout the heat. Most furnace operations will be fully automated. Charging of scrap into the preheater will be fully automated based on the scrap height in the chamber. Carbon and oxygen injection will be controlled based on the depth of foamy slag.

The results presented in Table 5.15 have been reported for the IHI shaft furnace.

Table 5.15 Reported results at Tokyo Steel

Items	Utonomiya plant	Takamatsu plant
Tap-to-tap time	60min	45min
Poweron time	55min	40min
Power consumption	236kW·h/t liquid	236kW·h/t liquid
Electrode consumption	1.0kg/t liquid	1.0kg/t liquid
Oxygen consumption	28Nm^3/t liquid	25Nm^3/t liquid
Carbon consumption	27kg/t liquid	25kg/t liquid

5.5.9.2 VAI Comelt

VAI has been working on a new furnace design they call Comelt. This furnace features a shaft and four individually moveable graphite electrodes which protrude into the furnace at an angle. A bottom anode is installed in the hearth center. The power supply is DC. The lower part of the shaft contains openings for lime and coke additions via a bin system. The upper part of the shaft contains a lateral door for scrap charging and an opening through which the furnace offgases flow. Scrap is charged to the shaft via a conveyor belt.

The furnace consists of a tiltable furnace vessel and a fixed shaft which are connected with a movable shaft ring. The furnace employs eccentric bottom tapping. The whole vessel structure rests on a tiltable support frame. The upper part of the vessel, the shaft and shaft ring are all lined with water-cooled panels. The shaft is enclosed with a steel structure which sits on a carriage. The shaft can be moved off the furnace shell via this carriage. The shaft ring provides a tight connection between the furnace vessel and the shaft. Key furnace components are shown in Fig. 5.36.

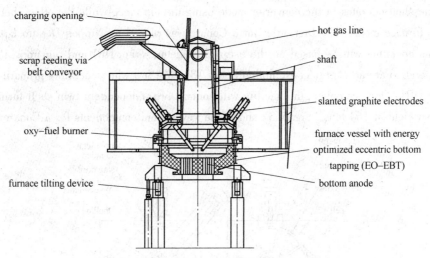

Fig. 5.36 Key components of the VAI Comelt furnace

Following tapping, the furnace is charged through the shaft with up to 80% of the total charge weight, lime and carbon. After charging, the charging door in the top of the shaft is closed and the electrodes are moved into working position. Oxy-fuel door burners are used to clear an area for insertion of oxygen lances which immediately begin to inject oxygen into the heel. Additional lances higher up in the furnace provide post-combustion oxygen. As the scrap melts in and the shaft begins to empty, additional scrap is fed to the shaft. The electrode arrangement allows the flow of gases to penetrate down into the scrap and then rise up through the scrap column thus maximizing heat recovery.

VAI has projected that the capital cost for Comelt will be greater than that for an equivalent DC operation for small to medium heat sizes. This is due to the large investment required for the electrical systems for the electrodes. For large heat sizes however, the projections indicate that Comelt should be less expensive than conventional DC technology.

Though this process is still in the pilot plant stage, projections indicate that this furnace will have lower power (304kW · h/t liquid) and electrode (0.8kg/t) consumption than either conventional AC or DC furnaces.

5.5.9.3 Mannesmann Demag Huettentechnik Conarc

The MDH Conarc is based on an operation which combines converter and EAF technologies, namely a converter arc furnace. This technology is based on the growing use of hot metal in the EAF and is aimed at optimizing energy recovery and maximizing productivity in such an operation. The use of hot metal in the EAF is limited by the maximum oxygen blowing rates which are dependent in turn on furnace size. The basic concept of Conarc is to carry out decarburization in one vessel and electric melting in another vessel. The Conarc system consists of two furnace shells, one slewable electrode structure serving both shells, one electric power supply for both shells, and one slewable top oxygen lance serving both shells.

Thus one shell operates in the converter mode using the top lance while the other shell operates in the arc furnace mode. The first order for a Conarc was placed by Nippon Dendro Ispat, Dolvi, India. This operation will be based on the use of hot metal, scrap, DRI and pig iron. Projections for this operation are an energy consumption below $181 kW \cdot h/t$ when operating primarily with hot metal and DRI. Once complete, this facility will contain two independent twin shell furnaces each with a tap weight of 164 tons. Fig. 5.37 shows the key system components for a Conarc facility.

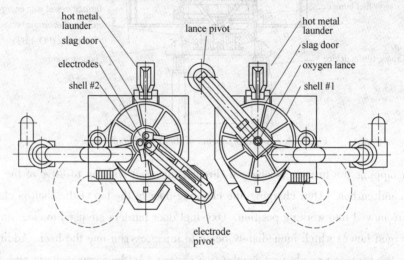

Fig. 5.37 Schematic of the Conarc process components

A Conarc has also been ordered for the Saldahna Steel facility which is being built in Saldahna Bay, South Africa. This operation will be based on 45% Corex hot metal and 55% DRI feed to the furnaces. Hot metal will be charged to one shell and the top lance will be used for decarburization. Simultaneously, DRI will be added to recover the heat generated. Once the aim carbon level is achieved, the electrodes will be moved to this shell and additional DRI will be fed to make up the balance of the heat.

Large quantities of heat are generated during the oxygen blowing cycle. As a result it is important that charge materials be added during this period so that some of this energy can be recovered. This will also help to protect the furnace shell from overheating.

5.5.9.4 Mannesmann Demag Huettentechnik Contiarc

The Contiarc EAF is a technology that has been proposed by MDH. It is based on a stationary ring shaft furnace with a centrally located DC arc heating system. Scrap is fed continuously by conveyor to the ring shaft. There the scrap is picked up by a series of magnets and is distributed evenly throughout the ring shaft area. The magnet train runs on a circular track below the furnace roof. Fig. 5.38 shows the main components of the Contiarc.

The scrap settles in the ring shaft and is preheated as furnace offgas rises up through the furnace. The descending scrap column is monitored using a level measuring system. If the charge sinks unevenly, additional scrap is fed to the lower points to restore an even scrap height profile.

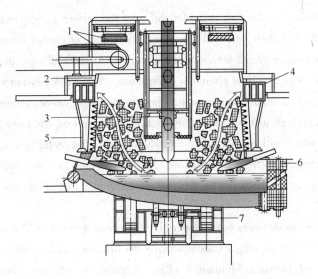

Fig. 5.38 Elements of the Contiarc process
1—scrap feeding and distribution system; 2—inner and outer ring type shell;
3—graphite electrode with guiding and lifting device; 4—offgas pipe;
5—fume hood; 6—syphon tapping system; 7—bottom electrode

Scrap is always present to protect the furnace walls so the arc can run continuously at maximum power without fear of damage to the furnace sidewalls.

The graphite electrode is located in a protective inner vessel to protect it from falling scrap. The inner vessel has a wear guard and a cooling water system in the lower hot zone. An electrically insulated ceramic electrode bushing is provided at the point where the electrode penetrates out into the furnace. Electrode regulation is performed hydraulically.

Chemical energy input, such as burners, is highly efficient because gas residence times in the Contiarc are greater than for conventional furnaces. Burners are located near the taphole to superheat the steel prior to tapping. Burners are also located near the slag door to facilitate slagging off. Slag free tapping is performed intermittently using a syphon tapping system.

Energy efficiency is extremely high using Contiarc as heat losses to the shell are minimized due to the continuous cover provided by the scrap. The elimination of top charging also reduces heat losses. The furnace is essentially airtight so that all offgas rises up through the scrap and is collected at the ring header at the top of the shaft. Dust losses are reduced 40% over conventional EAF operations as the scrap acts as a filter, trapping the dust as the offgas rises through the scrap. MDH projections indicate that the total energy input requirement for Contiarc will be only 62% of that required by an equivalent conventional EAF operation.

5.5.9.5 Future Melting Furnace Design

Without doubt, current trends in EAF design indicate that high levels of both electrical and chemical energy are likely to be employed in future furnace designs. As so many have pointed out, we are headed towards the oxy-electric furnace. The degree to which one form of energy is used over another will be dependent on the cost and availability of the various energy forms in a particular location. Raw materials and their cost will also affect the choice of energy source. The use of alternative iron sources containing high levels of carbon will necessitate the use of high oxygen blowing rates which equate to high levels of chemical energy use. A word of caution is in order though for such operations, as some thought must enter into how to maximize energy recovery from the offgases generated. In addition, high levels of materials such as cold pig iron can sometimes lead to extended tap-to-tap times which will reduce furnace productivity. If hot metal is used, some method must be provided to recover energy from the offgases. Materials which are also high in silicon will provide additional energy to the bath but at the cost of additional flux requirements and greater slag quantities generated.

A closed furnace design appears to be imperative for complete control of reactions in the furnace freeboard. Minimization of the offgas volume exiting the furnace will help to minimize offgas system requirements. If air infiltration is eliminated, offgas temperatures will be quite high which will give more efficient heat transfer in scrap preheating operations. If high levels of hydrogen and CO are present in the cooled gas, it could be used as a low grade fuel. Alternatively, if the CO is postcombusted in several preheater stages, maximum energy recovery to the scrap could be achieved.

The main disadvantage to operation of a closed furnace is that conventional oxygen lancing and coal injection operations would be difficult to carry out. These can however be carried out in a closed furnace if a series of submerged and side injectors are installed in the furnace shell and sidewalls similar to those used in conventional BOFs. The required injection rates would likely require that the furnace bath depth be increased in order to minimize blow through. A deeper bath with intense mixing will be beneficial for slag bath mixing which should improve steel quality and maximize recovery of iron units. Generation of CO in the bath will help to flush nitrogen and hydrogen out of the bath.

Some form of scrap preheating should be utilized to recover heat from the offgas. The best method for scrap preheating is still under debate but staged preheating coupled with post-combustion in each stage could help to minimize the need for a secondary post-combustion system downstream in the offgas system.

If a slanted electrode configuration similar to that used in Comelt is employed, it would be conceivable to retract the electrodes if necessary during part of the tap-to-tap cycle. This would allow preheated scrap to be charged periodically if a finger type preheat shaft were used. Alternatively the scrap could be continuously charged into a shaft by conveyor or a preheat tunnel conveyor similar to that used in Consteel, could be installed. A DC arc could be directed at the point where scrap feed falls into the furnace similar to the IHI furnace concept. This would help to maximize heat transfer from the arc to the scrap. If a scrap buildup on the walls of the furnace can be maintained, heat losses to the furnace shell can also be minimized.

5.5.9.6 Conclusions

There are many new processes for steelmaking which are now being commercialized. In almost all cases the goal is to minimize the electrical energy input and to maximize the energy efficiency in the process. Thus several technologies have attempted to maximize the use of chemical energy into the process (EOF, K-ES, LSF, etc.). These processes are highly dependent on achieving a pseudoequilibrium where oxygen has completely reacted with fuel components (carbon, CO, natural gas, etc.) to give the maximum achievable energy input to the process. Other processes have attempted to maximize the use of the energy that is input to the furnace by recovering energy in the offgases (Fuchs shaft furnace, Consteel, EOF, IHI Shaft). These processes are highly dependent on good heat transfer from the offgas to the scrap. This requires that the scrap and the offgas contact each other in an optimal way.

All of these processes have been able to demonstrate some benefits. The key is to develop a process that will show process and environmental benefits without having a high degree of complexity and without affecting productivity. There is no perfect solution that will meet the needs of all steelmaking operations. Rather, steelmakers must prioritize their objectives and then match these to the attributes of various furnace designs. It is important to maintain focus on the following criteria:

1. To provide process flexibility.
2. To increase productivity while improving energy efficiency.
3. To improve the quality of the finished product.
4. To meet environmental requirements at a minimum cost.

With these factors in mind, the following conclusions are drawn:

1. The correct furnace selection will be one that meets the specific requirements of the individual facility. Factors entering into the decision will likely include availability of raw materials, availability and cost of energy sources, desired product mix, level of post furnace treatment/refining available, capital cost and availability of a trained workforce.

2. Various forms of energy input should be balanced in order to give the operation the maximum amount of flexibility. This will help to minimize energy costs in the long run, i.e. the capability of running with high electrical input and low oxygen or the converse.

3. Energy input into the furnace needs to be well distributed in order to minimize total energy requirements. Good mixing of the bath will help to achieve this goal.

4. Oxygen injection should be distributed evenly throughout the tap-to-tap cycle in order to minimize fluctuations in offgas temperature and composition. Thus postcombustion operations can be optimized and the size of the offgas system can be minimized. In addition, fume generation will be minimized and slag/bath approach to equilibrium will be greater.

5. Injection of solids into the bath and into the slag layer should be distributed across the bath surface in order to maximize the efficiency of slag foaming operations. This will also enable the slag and bath to move closer to equilibrium. This in turn will help to minimize flux requirements and will improve the quality of the steel.

6. Submerged injection of both gases and solids should be maximized so that the beneficial effects of bath stirring can be realized. Slag fluxes could be injected along with oxygen. The high temperatures achieved in the fireball coupled with the high oxygen potential will help the lime flux

into the slag faster, and as a result, dephosphorization and desulfurization operations can proceed more optimally.

7. If injection of solids and gases is increased, it will likely be beneficial to increase the bath depth. This will also be beneficial for steel quality. For operations using high levels of solids injection (those feeding high levels of iron carbide), a deeper bath will help to reduce blow through (solids exiting the bath) and could also allow higher injection rates.

8. The melting vessel should be closed up as much as possible in order to minimize the amount of air infiltration. This will minimize the volume of offgas exiting the furnace leading to smaller fume system requirements. In the case of post-combustion operations coupled with scrap preheating, the gas volume will be minimized while maximizing offgas temperature for efficient heat transfer to the scrap. If secondary post-combustion is required it can be achieved most effectively at minimum cost by minimizing the volume of offgas to be treated.

9. If high carbon alternative iron sources are used as feed in scrap melting operations, some form of post-combustion is imperative in order to recover energy from the high levels of CO contained in the offgas. For these operations, energy recovery will likely be maximized by coupling post-combustion with some form of scrap preheating.

10. Scrap preheating provides the most likely option for heat recovery from the offgas. For processes using a high degree of chemical energy in the furnace, this becomes even more important, as more energy is contained in the offgas for these operations. In order to maximize recovery of chemical energy contained in the offgas, it will be necessary to perform post-combustion. Achieving high post-combustion efficiencies throughout the heat will be difficult. Staged post-combustion in scrap preheat operations could optimize heat recovery further.

11. Alternatively, operations which generate high levels of hydrogen and carbon monoxide may find it cost effective to try to recover the calorific energy contained in the offgas in the same manner as some BOF operations which cool and clean the gas prior to using it as a low grade fuel. For this type of operation, it will be necessary to operate with a closed furnace. To minimize fuel gas storage requirements, generation of CO in the furnace should be balanced throughout the cycle. This can be achieved by using side shell or bottom oxygen injectors and by injecting oxygen at constant levels throughout the heat. If this form of operation is coupled with scrap preheating, any VOCs resulting from the scrap will form part of the offgas stream. A greater scrap column height can be utilized because secondary post-combustion downstream in the offgas system is not necessary. The high scrap column will also help strip the fume and dust from the offgas stream thus minimizing cleaning requirements. The resulting fume system requirement will be considerably smaller than that for a conventional furnace with similar melting capacity.

12. In the future, it will likely be necessary to recycle furnace dust and mill scale back into the steelmaking vessel. Preliminary trials indicate that this material is a good slag foaming agent. Key drawbacks are handling the dust prior to recycle and its concentration buildup of zinc and lead over time.

13. Twin shell furnace configurations will continue to be used in order to reduce cycle times be-

low 45 minutes (comparable to BOF operations) and to optimize power-on time. In the case of operations using a high percentage of hot metal, operations similar to Conarc will be used, thus employing oxygen blowing into a hot heel with scrap addition. Thought must be given to maximizing recovery of energy from the hot offgases generated during periods of high oxygen injection rates.

14. The use of hot metal in the EAF will increase as more high tonnage EAF operations attempt to achieve cycle times of 35-40 minutes. Hot metal should be added to the heat gradually so that large fluctuations in the bath chemistry do not occur which could result in explosions. In addition, high decarburization efficiency can be achieved if the bath carbon level is kept at approximately 0.3% until final refining takes place (blowing down the desired tap carbon level at the end of the heat). Decarburizing from high carbon levels has been shown to generate high iron losses to fuming. At very low bath carbon levels, large amounts of FeO are generated and report to the slag.

15. DC furnaces will continue to be of interest for high efficiency melting operations. The additional space available on the furnace roof due to a single electrode operation will allow for much greater automation of operating functions such as bath sampling, temperature sampling, solids/gas injection, etc. The concept of using side slanted electrodes as in the case of Comelt affords several advantages to furnace operation and could allow for electrodes to be retracted in some phases of the tap-to-tap period. This would allow scrap to be charged from a preheating shaft without fear of electrode damage. The use of DC furnaces might also allow for smaller furnace diameters coupled with greater bath depth.

16. Operations which desire maximum flexibility at minimum cost will result in more hybrid furnace designs. These designs will take into account flexibility in feed materials and will continue to aim for high energy efficiency coupled with high productivity. For example operations with high solids injection, iron carbide or DRI fines, may choose designs which would increase the flat bath period in order to spread out the solids injection cycle. Alternatively, a deeper bath may be used so that higher injection rates can be used without risk of blow through.

17. Operating practices will continue to evolve and will not only seek to optimize energy efficiency in the EAF but will seek to discover the overall optimum for the whole steelmaking facility. Universally, the most important factor is to optimize operating costs for the entire facility and not necessarily one operation in the overall process chain.

Along with added process flexibility comes greater process complexity. This in turn will require greater process understanding so that the process may be better controlled. Much more thought consequently must enter into the selection of electric furnace designs and it can be expected that many new designs will result in the years ahead. As long as there is electric furnace steelmaking, the optimal design will always be strived for.

Exercises

5-1 What should be controlled carefully for EAF steelmaking?

5-2 How long is the tap-to-tap time for EAF now and before?

5-3 Please expound the significance of the application of oxygen lancing in the EAF.

5-4 How to realize the bottom stirring for EAF?

5-5 What are the demands of EAF for raw materials?

5-6 What are the normal fluxes and additives for EAF?

5-7 Please expound the "typical 60 minutes tap-to-tap cycle" for EAF operating.

5-8 What about the composition for the typical slag of EAF?

5-9 Please show the different kinds of the new scrap melting processes.

References

[1] R. J. Fruehan. The AISE Steel Foundation [M]. America: PA, 1998.

[2] H. C. Van Ness. Understanding Thermodynamics (Dover Books on Physics) [M]. America: Dover Publications, 1983.

[3] Ahindra Ghosh, et al. Ironmaking and Steelmaking: Theory and Practice [M]. America: Prentice-Hall of India Pvt. Ltd, 2008.

[4] Dipak Mazumdar. A First Course in Iron and Steelmaking [M]. America: Orient Blackswan, 2015.

[5] 战东平，等．大学专业英语：材料英语 [M]．北京：外语教学与研究出版社，2002.

[6] 李进．冶金工程专业英语 [M]．北京：冶金工业出版社，2013.

[7] 陈家祥．钢铁冶金学：炼钢部分 [M]．北京：冶金工业出版社，2006.

[8] 王新华．钢铁冶金：炼钢学 [M]．北京：高等教育出版社，2007.

[9] 金焱．冶金工程专业英语 [M]．北京：冶金工业出版社，2008.

[10] 雷亚，等．炼钢学 [M]．北京：冶金工业出版社，2010.